Laboratory Manual For

Exploring Electronic Devices

By Mark Edward Hazen
Brevard Community College

AF853343

Saunders College Publishing,
a division of
Holt, Rinehart and Winston, Inc.

Saunders College Publishing
Philadelphia Fort Worth Chicago San Francisco
Montreal Toronto London Sydney Tokyo

To the memory of my grandfather, Arthur Dyre Gibbs, who inspired me at a young age with his electronic gadgets and short-wave radios.

Printed in the United States of America

International Standard Book Number: 0-03-028534-8

67 017 9876543

Laboratory Manual for Exploring Electronic Devices

PREFACE

This laboratory manual is designed for a two-semester course in electronic devices and circuits and is intended to coincide with *Exploring Electronic Devices* . As can be seen from the table of contents, an attempt has been made to provide a balance between the number of discrete-device experiments and integrated-circuit experiments. At the beginning of each experiment, in the Preparations section, the experiment is referenced directly to the text. This enables you to read the background information relating to the subjects of the experiment.

All data collected during the experiment should be placed in tables provided on the answer sheets at the end of each experiment. The answer sheets can be removed and submitted for grading.

If you desire, some of the experiments can be used to construct very useful pieces of equipment. Components used in, and knowledge gained from, Experiments 3, 4, 5, 41, 42, and 43 can be used to construct a regulated DC power supply for your shop. Experiment 37 can be used to construct a function generator. In time, I am sure you will find many practical uses for the op-amp mixers and filters and the 555 timer IC.

Finally, many helpful appendices have been provided at the end of this manual. You should take time to familiarize yourself with them. Enjoy your explorations!

Mark E. Hazen

- IMPORTANT NOTICE -

In many of the experiments in this manual, a 115 V power transformer is used as part of the circuits. This transformer must be prepared carefully so it can be used safely. Care must be taken to properly secure and insulate the power cord connections to the transformer's primary leads in order to prevent accidental electrical shock. The author highly recommends that this transformer be placed in an insulated box with banana terminals for the low-voltage secondary leads. Due to the fact that the author and publisher have no control over the preparation and safe use of these transformers, neither the author nor the publisher can be held responsible for any injury or damage to persons or property that may result from their use while conducting these experiments or not. It is therefore the responsibility of the institution and/or instructor to insure the safety of persons and property in regard to these transformers and any experiments conducted by any persons.

Laboratory Manual for Exploring Electronic Devices

TABLE OF CONTENTS

APPENDICES

Diode Characteristics and Testing

EXPERIMENT 1

PURPOSE

The purpose of this experiment is to familiarize you with the forward- and reverse-bias characteristics of semiconductor diodes.

FORMAL OBJECTIVES

After completing this experiment, you will be able to:

- use the ohmmeter or diode-test function of a multimeter to test a diode out-of-circuit.
- compare real diode characteristics to ideal characteristics.
- explain the effect forward and reverse bias has on diode current and static resistance.

PREPARATIONS

In preparation for this experiment, you should study Sections 1-3, 2-1, and 2-2 of *Exploring Electronic Devices*. Also, take the time to read through this experiment before entering the lab to actually perform the experiment.

COMPONENTS and EQUIPMENT

1 - Variable DC source (0 to 15 V, 250 mA capability)
1 - Multimeter (Analog or Digital)
1 - 10 Ω, 1/4 W Resistor
1 - 100 Ω, 1/4 W Resistor
1 - 1N914 Switching Diode
1 - Protoboard (Plug-in Experiment Board)
1 - 100 Ω, 1 W Resistor
1 - 1 kΩ, 1/4 W Resistor
1 - 1N4001 Rectifier Diode

PART ONE: USING THE MULTIMETER TO TEST A DIODE

In this part of the experiment, you will use your multimeter to test your diodes to see if they are good. Before you begin, let's discuss what you should expect to see. As you know from your reading, the diode acts like an automatic one-way valve. The diode conducts significant amounts of current only in one direction, when the diode is forward biased - negative potential on the cathode (marked end) and positive potential on the anode. Here, you will use your multimeter to forward and reverse bias the diode to see if it is good. The internal battery of your multimeter will supply the bias voltage.

Using an Analog Ohmmeter: If you are using an analog ohmmeter, you will want to use a second voltmeter to determine the actual polarity of the ohmmeter's test leads. Very often, the red test lead is the negative potential and the black lead is the positive potential for analog multimeters set to the ohms function. Use a low ohms range to test your diodes.

Using a Digital Ohmmeter: If you are using a digital multimeter in the ohms function, set the range selector to a high ohms range such as 10 or 20 MΩ to test your diodes. For digital ohmmeters, the red lead is usually the positive potential and the black lead is the negative potential, as you might expect.

med = 623 infinite

1. Use your ohmmeter as described above to forward bias the 1N4001 diode. Record the type of meter used, the range setting, and the reading.

   Meter Type: Fluke ______ Range: 2k ______ Reading: 642 Ω ______

small = 687 infinite

2. Use your ohmmeter to reverse bias the 1N4001 diode. Record the reading.

   Reading with reverse bias: infinite ______

3. Use your ohmmeter to forward bias the 1N914 diode. Record the reading.

   Reading: 568 Ω ______

4. Use your ohmmeter to reverse bias the 1N914 diode. Record the reading.

   Reading with reverse bias: infinite ______

Simpson meter

5. Diode Test Function: If your multimeter has the capability, use the diode-test function to test your diodes. When this function is used, a voltage is applied to the diode under test through an internal series resistance. The meter actually reads the voltage drop across the diode that is being tested. Thus, a forward reading is the actual forward-biased junction voltage of the diode.

   1N4001 Forward Test Reading: 21 Ω ______, Reverse Reading: 300k Ω ______

   1N914 Forward Test Reading: 18 Ω ______, Reverse Reading: infinite ______

6. Answer all questions for Part One on the answer sheet.

PART TWO: FORWARD-BIAS DIODE CHARACTERISTICS

In this part of the experiment, you will collect forward current and voltage data for the 1N4001 rectifier diode and the 1N914 switching diode. You will then use this data to establish a graph of current vs voltage for each diode. Finally, you will analyze the data and graphs to identify important real characteristics of forward-biased diodes. We will not perform reverse-bias current measurements because the reverse current is usually far less than most multimeters can measure.

Figure 1

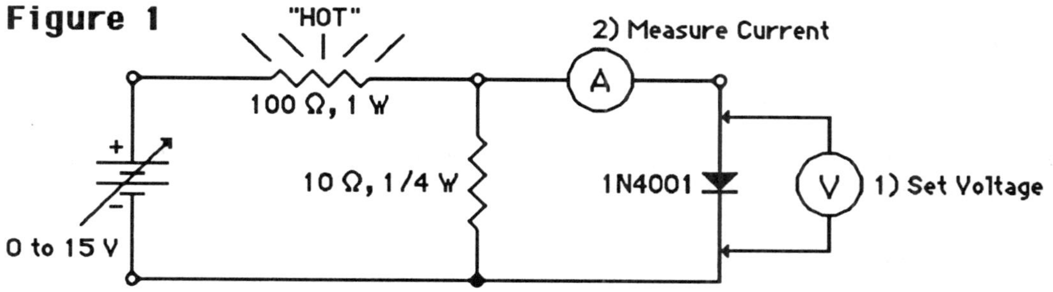

1. Assemble the circuit shown in Figure 1. Preset the DC voltage control to 0 V. DC supply should be turned off.

2. Note that your multimeter is being used as an ammeter to measure forward diode current (I_F) and a voltmeter to measure forward diode voltage (V_F). If a second meter is available, connect it in series with the diode to measure current as shown. If a second meter is not available, you will have to first set the diode voltage, then insert the meter as an ammeter to measure diode current.

3. Apply power to the DC supply. With your voltmeter across the diode, slowly increase the DC voltage control so the diode voltage increases in steps as indicated in Table 1 on the answer sheet. At each voltage step, record the forward diode current.

   CAUTION: THE 100-Ω SERIES RESISTOR MAY BECOME HOT!

4. Turn off the DC supply and return the DC voltage control to 0 V.

5. When all readings have been recorded, calculate the static forward resistance of the diode for each value of forward voltage and current. ($R_{static} = V_F/I_F$) Record your answers in Table 1 in the column provided.

6. Next, plot your current and voltage data to form a forward current and voltage characteristic curve on Chart 1.

Figure 2

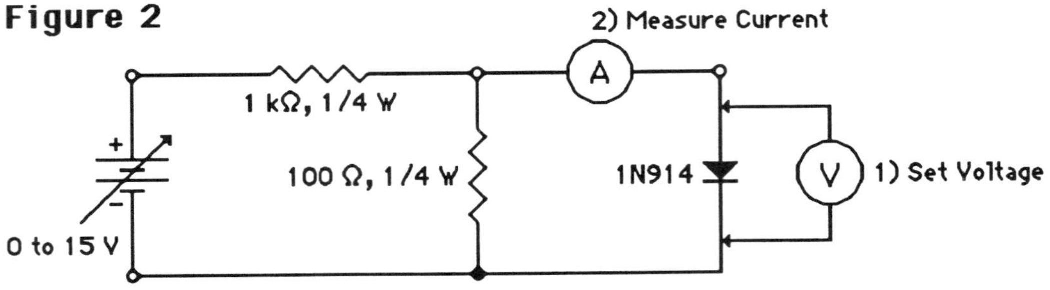

7. Replace the 100-Ω series resistor with a 1-kΩ resistor, the 10-Ω resistor with a 100-Ω resistor, and the 1N4001 with a 1N914 as shown in Figure 2.

8. Repeat the experiment again recording data in Table 2 and Chart 2.

9. Answer all questions for Part Two.

NAME:______________________ CLASS:__________ DATE:__________

PART ONE QUESTIONS

1. In the space provided here, draw a diagram showing how you used your multimeter (ohmmeter) to test your diodes in the forward- and reverse-biased modes. Label the polarity of your test leads and the diode terminals.

Forward-Biased Mode   Reverse-Biased Mode

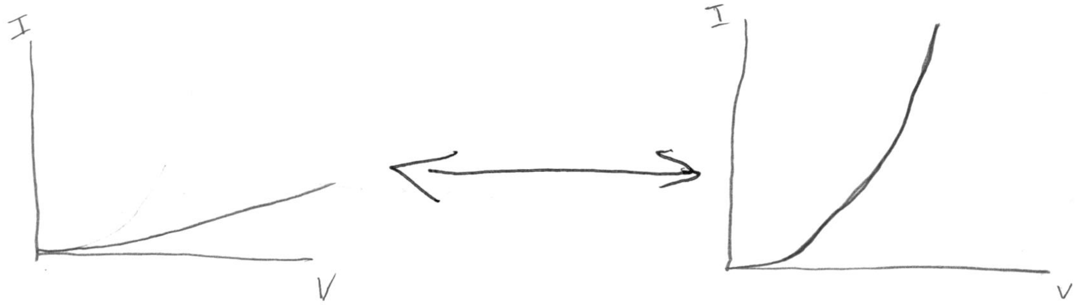

2. Describe the differences you observed between the forward and reverse-bias readings of your diodes when you tested them using the ohmmeter.
More Current passed through the reversed Biased Mode at the same voltage levels

3. When the diode test function of your multimeter is used and the diode is forward biased what is the significance of the reading you get (what is the reading)?
The reverse Biased mode had more resistance

4. According to the data recorded in Step 5, do you think all diodes have the exact same forward junction voltage? No

5. If a diode is shorted, what general readings would you expect to see while testing the diode with your ohmmeter?
no resistance

6. If a diode is open, what general readings would you expect to see while testing the diode with your ohmmeter?
infinite resistance

PART TWO QUESTIONS

Table 1

V_F	I_F	R_{static}
0.1	0	
0.2	1.6 μA	.125 kΩ
0.3	2.5	.12 kΩ
0.4	3.3	.12 kΩ
0.5	4.4	.11 kΩ
0.55	5.6	.10 kΩ
0.6	7.9	.09 kΩ
0.65	16.3	.05 kΩ
0.7	49.2	.018 kΩ
0.75	110	.09 kΩ

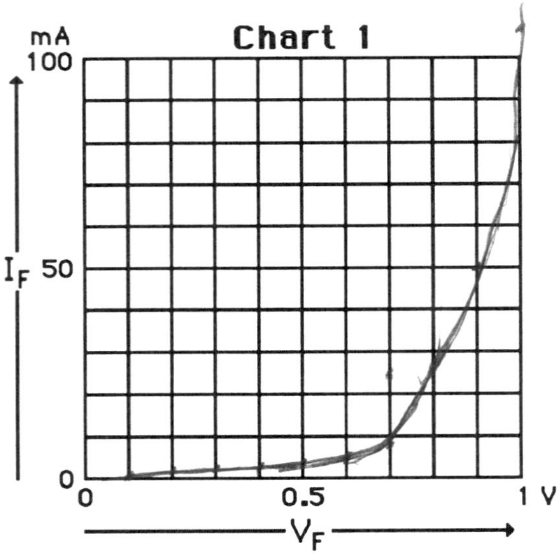

Table 2

V_F	I_F	R_{static}
0.1	.8 μA	.125
0.2	1.3 μA	1.5
0.3	2.8	.107
0.4	3.4	.118
0.5	4.6 μA	.108
0.55	5.6 μA	.107
0.6	6.6	.106
0.65	7.6	.105
0.7	8.7	.103
0.75	9.85	.010

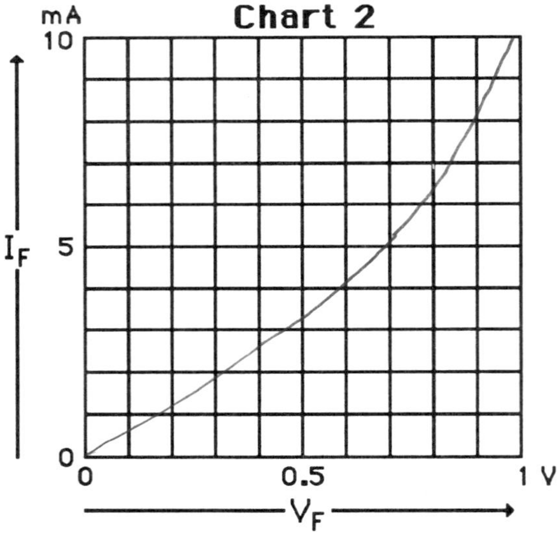

1. At what value of forward voltage ($V_{F(on)}$) did the 1N4001 seem to really turn on where the forward current began to increase significantly? .6V

2. At what value of forward voltage ($V_{F(on)}$) did the 1N914 seem to really turn on where the forward current began to increase significantly? .7v

3. Which of the following do the characteristic curves of Chart 1 and Chart 2 indicate?

   a) forward diode current increases at an infinite rate for forward-bias voltages greater than approximately 0.6 V.
   b) forward diode current increases at a fairly steep rate for forward-bias voltages greater than approximately 0.6 V.
   c) forward diode current increases at a gradual rate for forward-bias voltages greater than approximately 0.6 V.

4. From your reading in Exploring Electronic Devices, why is there a forward voltage drop across the diode when it is forward biased?
   Because

5. Why does the forward diode voltage increase slightly with a fairly large increase in forward current after $V_{F(on)}$ has been reached?
   Because the resistance is becoming less

6. If the diodes were ideal, with $V_{F(on)}$ being considered, how would the characteristic curves look in Chart 1 and Chart 2? Use a second color to draw these in Chart 1 and Chart 2.

7. Describe the forward- and reverse-bias characteristics of a real semiconductor diode.
   When the diode is forward bias the resistance is lower and current is allowed to pass through. Reverse bias have infinite resistance and the circuit cannot be completed

NOTES AND CALCULATIONS

NOTES AND CALCULATIONS

Half-Wave Rectifier Circuits

EXPERIMENT 2

PURPOSE

The purpose of this experiment is to acquaint you with the theory and operation of half-wave rectifier circuits using a single semiconductor diode. This experiment will also highlight basic troubleshooting using the oscilloscope.

FORMAL OBJECTIVES

After completing this experiment, you will be able to:

- identify half-wave rectifier circuits as either positive or negative half-wave rectifiers.
- calculate peak pulsating DC voltage and current from a half-wave rectifier when the amplitude of the applied AC is known and the load resistance is known.
- calculate average DC voltage and current from a half-wave rectifier when the amplitude of the applied AC is known and the load resistance is known.
- troubleshoot half-wave rectifier AC to DC converters.

PREPARATIONS

In preparation for this experiment, you should study Section 2-3 of *Exploring Electronic Devices*.

COMPONENTS and EQUIPMENT

1 - Dual-trace oscilloscope with test probes
1 - Multimeter (analog or digital)
1 - Transformer, 25.2 V, 450 mA, center-tapped secondary
1 - Power cord, approximately 6 ft., three wire with three-prong plug
1 - 1N4001 Diode
1 - Protoboard
1 - 470 Ω, 1/4 W Resistor

PART ONE: THE POSITIVE HALF-WAVE RECTIFIER

In this part of the experiment, you will assemble and test a positive half-wave rectifier circuit. You will use the oscilloscope to view and measure the applied AC waveform and the rectified pulsating DC waveform across the load resistance. You will use this information to calculate average load voltage, current, and power dissipation where,

$V_{P(load)} \sim V_{P(source)} - 0.7\text{ V}$ $V_{avg} = 0.318 V_{P(load)}$ $I_{avg} = V_{avg}/R_L$ $P_D = I_{avg} \bullet V_{avg}$

1. The first task at hand is to prepare the transformer for use in the experiment. The power cord has three wires, one is the hot wire (probably black), one is the neutral wire (probably white), and the third is the ground wire connected to the cylindrical ground prong on the plug (probably green). If you have any doubts, use your ohmmeter to check which wire is connected to which prong on the plug. Solder the hot and neutral wires to the primary side wires of your transformer as shown in Figure 1. Wrap the solder connections safely with electrical tape. Use a small bolt with nut and washers to secure the ground wire to one of the mounting tabs on the transformer. In the remote event that the primary winding shorts to the transformer core, the high voltage will be shorted to ground, trip a circuit breaker, and save your life.

   HAVE YOUR INSTRUCTOR CHECK YOUR WIRING BEFORE YOU CONTINUE!

Figure 1

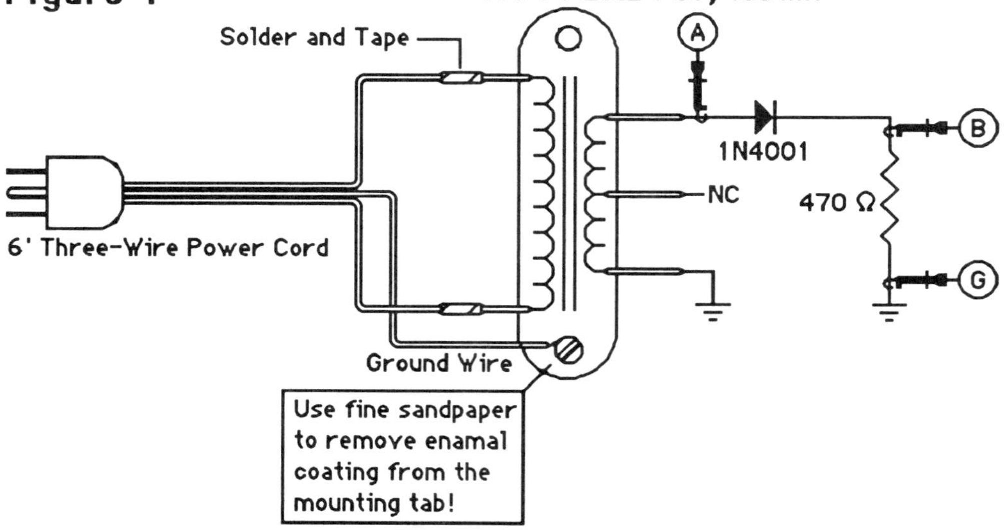

2. Finish assembling the circuit on the secondary side of the transformer as shown in Figure 1. Use clip leads or hook-up wire and a protoboard to complete the circuit. Notice the center tap on the secondary winding is not used (NC = no connection). The other secondary wire is used as common ground. The small clip-on probe symbols indicate test points in the circuit.

3. Power-on the oscilloscope and connect a probe cable to Channel 1. Connect the Channel 1 probe to testpoint A with ground clip to G. Set the oscilloscope to trigger on Channel 1.

4. Plug in the power cord for the transformer. Adjust the time/division selector so approximately two complete AC cycles are shown on the screen. Set the volts/division selector and the vertical position control so the AC cycles occupy only the upper half of the screen. The horizontal line in the center of the upper half of the screen is to be used as the zero-reference line.

5. Connect a probe cable to the Channel 2 input on the oscilloscope. Connect the Channel 2 probe to testpoint B with ground clip to G.

6. Set the Channel 2 volts/division selector and the vertical position control so the pulsating DC cycles occupy only the lower half of the screen. The volts/division controls for both channels should be set the same. The bottom horizontal line on the screen is to be used as the zero-reference line.

7. On the answer sheet in Part One, draw the waveforms that you see on Screen 1 and fill in the requested data.

8. Unplug your transformer, turn off the oscilloscope, and answer all questions for Part One. Do not disassemble your circuit.

PART TWO: THE NEGATIVE HALF-WAVE RECTIFIER

In this part of the experiment, you will investigate a negative half-wave rectifier circuit.

Figure 2

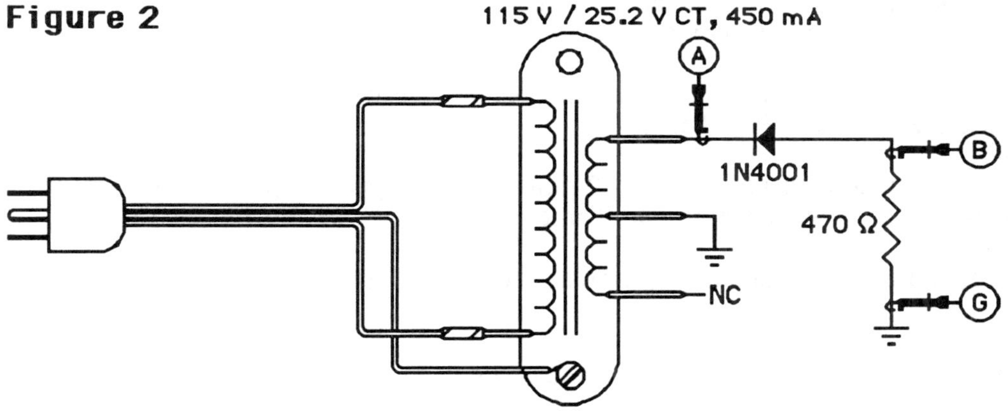

1. Reverse the diode as shown in Figure 2. Use the center tap as the common ground.

2. Power-on the oscilloscope and connect a probe cable to Channel 1. Connect the Channel 1 probe to testpoint A with ground clip to G. Set the oscilloscope to trigger on Channel 1.

3. Plug in the power cord for the transformer. Adjust the time/division selector so approximately two complete AC cycles are shown on the screen. Set the volts/division selector and the vertical position control so the AC cycles occupy only the upper half of the screen. The horizontal line in the center of the upper half of the screen is to be used as the zero-reference line.

4. Connect a probe cable to the Channel 2 input on the oscilloscope. Connect the Channel 2 probe to testpoint B with ground clip to G.

5. Set the Channel 2 volts/division selector and the vertical position control so the pulsating DC cycles occupy only the lower half of the screen. The volts/division controls for both channels should be set the same. The center horizontal line on the screen is to be used as the zero-reference line for the pulsating DC.

6. On the answer sheet in Part Two, draw the waveforms that you see on Screen 2 and fill in the requested data.

7. Unplug your transformer, turn off the oscilloscope, and answer all questions for Part Two. Do not disassemble your circuit.

PART THREE: TROUBLESHOOTING THE HALF-WAVE RECTIFIER

In this part of the experiment, you will troubleshoot the half-wave AC to DC converter. Various circuit troubles will be simulated and examined.

1. Plug in the power cord for the transformer and apply power to the oscilloscope.

Open Transformer Winding

2. Using the circuit of Figure 2, disconnect the center tap of the transformer secondary from the common ground. This will simulate an open secondary winding. It also simulates an open primary or open fuse in the primary circuit.

3. Use the oscilloscope to measure the peak load voltage at testpoint B. Record this measurement on the answer sheet in Part Three.

4. Use the oscilloscope to measure the peak-to-peak AC voltage at testpoint A. Record this measurement on the answer sheet in Part Three.

5. Restore the center tap to ground connection as shown in Figure 2.

Open Diode

6. Remove the diode. This will simulate an open diode.

7. Use the oscilloscope to measure the peak load voltage at testpoint B. Record this measurement on the answer sheet in Part Three.

8. Use the oscilloscope to measure the peak-to-peak AC voltage at testpoint A. Record this measurement on the answer sheet in Part Three.

9. Place the diode back in the circuit exactly as shown in Figure 2.

Shorted Diode

10. Place a short jumper wire across the diode to simulate a shorted diode.

11. Use the oscilloscope to measure the load voltage at testpoint B. Record this measurement on the answer sheet in Part Three.

12. Use the oscilloscope to measure the peak-to-peak AC voltage at testpoint A. Record this measurement on the answer sheet in Part Three.

13. Remove the jumper from across the diode.

14. Turn off the oscilloscope. Unplug the power cord for the transformer. Disassemble your circuit.

NOTES AND CALCULATIONS

NAME:____________________________ CLASS:__________ DATE:__________

PART ONE QUESTIONS

Screen 1

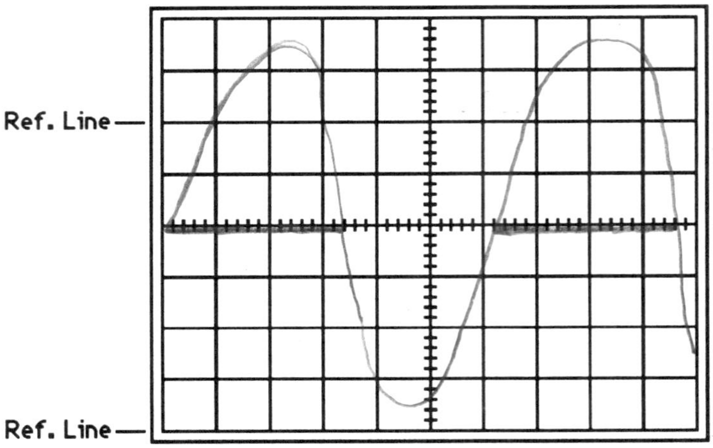

Oscilloscope Control Settings:
Time/Div: 5ms
—Volts/Div: 10v/div
—Volts/Div: 10 v/div

1. What is the peak-to-peak AC voltage at testpoint A in Figure 1? 70v
2. What is the peak AC voltage at testpoint A in Figure 1? 35v
3. What is the value of peak pulsating DC voltage at testpoint B in Figure 1? -34.3v
4. Calculate the average DC load voltage. 11.13v
5. Calculate the average DC load (current). .024a
6. Calculate the power dissipated in the load resistor. 264mW
7. Which of the two AC cycles is passed to the load resistor, (a) the positive, or (b) the negative alternation?
8. Which of these best describes the circuit of Figure 1?
   a) a positive full-wave-rectifier AC to DC converter
   b) a negative full-wave-rectifier AC to DC converter
   c) a positive half-wave-rectifier AC to DC converter
   d) a negative half-wave-rectifier AC to DC converter

PART TWO QUESTIONS

Screen 2

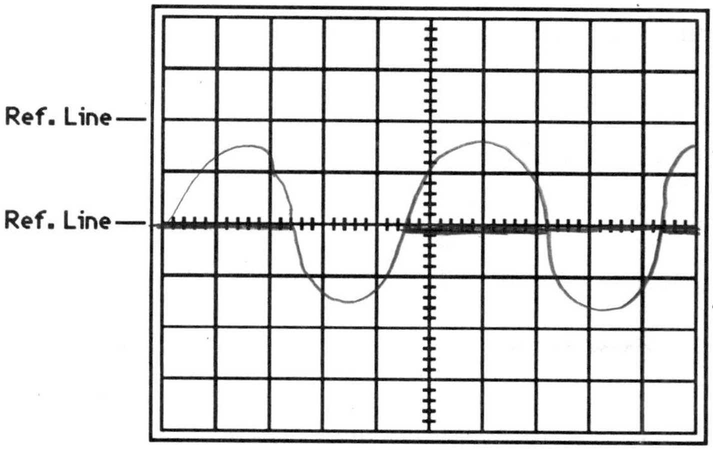

Oscilloscope Control Settings:
Time/Div : 5ms
Volts/Div : 5v/div

Volts/Div : 5v/div

1. What is the peak-to-peak AC voltage at testpoint A in Figure 2? 35v
2. What is the peak AC voltage at testpoint A in Figure 2? 17v
3. What is the value of peak pulsating DC voltage at testpoint B in Figure 2? -16.3v
4. Calculate the average DC load voltage. 5.406v
5. Calculate the average DC load (current). .011a
6. Calculate the power dissipated in the load resistor. .062mW
7. Which of the two AC cycles is passed to the load resistor, (a) the positive, or (b) the negative alternation?
8. Which of these best describes the circuit of Figure 2?
   a) a positive full-wave-rectifier AC to DC converter
   b) a negative full-wave-rectifier AC to DC converter
   c) a positive half-wave-rectifier AC to DC converter
   d) a negative half-wave-rectifier AC to DC converter
9. When looking at a schematic or an actual half-wave rectifier circuit, describe how to determine whether the half-wave rectifier provides a positive load voltage or a negative load voltage.
   By looking at which direction the diode is facing

PART THREE QUESTIONS

1. Load voltage with center tap disconnected from ground: 0

2. Secondary voltage at testpoint A with center tap disconnected from ground: 0

3. Load voltage with diode removed: .01v

4. Secondary voltage at testpoint A with diode removed: 0v

5. Load voltage with diode shorted: 12.52v

6. Secondary voltage at testpoint A with diode shorted: 17v

7. Describe the load voltage when the diode is shorted. load voltage is the RMS voltage.

8. Describe the measurements you would expect to see at testpoints A and B if the primary of the transformer were burned open or if a fuse in the primary circuit were burned open.
0v when burnt open

9. Fill in the possible-cause lines for Table 1.

Table 1

SYMPTOMS		POSSIBLE CAUSES
$V_{Secondary}$ Testpoint A	V_{Load} Testpoint B	
0 V	0 V	Lead Disconected or open diode
Normal AC Voltage	0 V	Open diode
Normal AC Voltage	Full AC Secondary V	Shorted diode
Normal AC Voltage	Normal Pulsating DC	Normal circuit

Full-Wave Rectifier Circuits

EXPERIMENT 3

PURPOSE

The purpose of this experiment is to acquaint you with the theory and operation of full-wave rectifier circuits. This experiment will also highlight some basic troubleshooting using the oscilloscope.

FORMAL OBJECTIVES

After completing this experiment, you will be able to:

- identify full-wave rectifier circuits as either positive or negative full-wave rectifiers.
- calculate peak pulsating DC voltage from a full-wave rectifier when the amplitude of the applied AC is known and the load resistance is known.
- calculate average DC voltage and current from a full-wave rectifier when the amplitude of the applied AC is known and the load resistance is known.

PREPARATIONS

In preparation for this experiment, you should study Section 2-4 of *Exploring Electronic Devices.*

COMPONENTS and EQUIPMENT

1 - Dual-trace oscilloscope with test probes
1 - Protoboard
1 - Multimeter (analog or digital)
1 - Transformer with attached power cord, 25.2 V, 450 mA, center-tapped secondary
4 - 1N4001 Diodes
1 - 470 Ω, 1/4 W Resistors

PART ONE: THE FULL-WAVE RECTIFIER USING A CENTER-TAPPED TRANSFORMER

In this part of the experiment, you will assemble and test a positive full-wave rectifier circuit that uses a center-tapped secondary and only two diodes. You will use the oscilloscope to view and measure the applied AC waveform and the rectified pulsating DC waveform across the load resistance. You will use this information to calculate average load voltage, current, and power dissipation where,

$V_{P(load)} \sim V_{P(source)} - 0.7\ V$ $\quad V_{avg} = 0.637V_{P(load)}$ $\quad I_{avg} = V_{avg}/R_L$ $\quad P_D = I_{avg} \bullet V_{avg}$

Using a center-tapped secondary, the peak source voltage for the rectifier circuit is half the total peak secondary voltage where,

$V_{P(sec)} = 1.414V_{sec(RMS)}$ and $V_{P(source)} = V_{P(sec)}/2$

1. Calculate the total peak secondary voltage and the peak source voltage for the 25.2 V center-tapped transformer you will use in this experiment. Record your answers on the answer sheet in Part One.

Figure 1

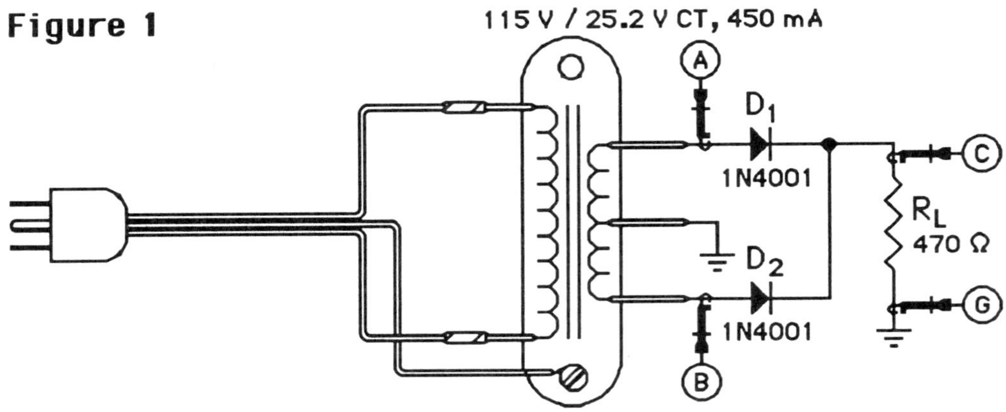

2. Assemble your full-wave rectifier circuit as shown in Figure 1. Use clip leads or hook-up wire and a protoboard to complete the circuit. The small clip-on probe symbols indicate test points in the circuit.

3. Power-on the oscilloscope and connect a probe cable to Channel 1. Connect the Channel 1 probe to Testpoint A with ground clip to G. Set the oscilloscope to trigger on Channel 1.

4. Plug in the power cord for the transformer. Adjust the time/division selector so approximately two complete AC cycles are shown on the screen. Set the volts/division selector and the vertical position control so the AC cycles occupy only the upper half of the screen. The horizontal line in the center of the upper half of the screen is to be used as the zero-reference line.

5. Connect a probe cable to the Channel 2 input on the oscilloscope. Connect the Channel 2 probe to Testpoint B with ground clip to G.

6. Set the Channel 2 volts/division selector and the vertical position control so the AC signal occupies only the lower half of the screen. The volts/division controls for both channels should be set the same. The horizontal line in the center of the lower half of the screen is to be used as the zero-reference line.

7. On the answer sheet in Part One, draw the waveforms that you see on Screen 1 and fill in the requested data.

8. Are the two signals displayed on the oscilloscope in phase with each other? ____________

9. Remove the Channel 2 probe from Testpoint B and place it at Testpoint C. Use the bottom line on the screen as the reference for the Testpoint-B signal.

10. On the answer sheet in Part One, draw the waveforms that you see on Screen 2 and fill in the requested data.

11. Remove D_1 from the circuit. Use the answer sheet to describe what you see on the oscilloscope and why.

12. Place D_1 back in the circuit.

Caution: In the following steps, be sure to observe polarity when inserting the ammeter in the circuit. Always remove power while connections are being made to avoid an accidental short. Make sure all connections are secure and correct before applying power.

13. Unplug your transformer. Use your multimeter as an ammeter and place it in series with the load resistor at Testpoint C. Once all connections are safely made, plug the power cord back in. Measure the current and record the reading on the answer sheet in Part One.

14. Unplug your transformer. Use your multimeter as an ammeter and place it in series with D_1 at Testpoint A. Once all connections are safely made, plug the power cord back in. Measure the current and record the reading on the answer sheet in Part One.

15. Unplug your transformer. Use your multimeter as an ammeter and place it in series with D_2 at Testpoint B. Once all connections are safely made, plug the power cord back in. Measure the current and record the reading on the answer sheet in Part One.

16. Unplug your transformer, turn off the oscilloscope, and answer all remaining questions for Part One.

PART TWO: THE BRIDGE RECTIFIER

In this part of the experiment, you will investigate a positive bridge rectifier circuit. Once again you will use the oscilloscope to view and measure the applied AC waveform and the rectified pulsating DC waveform across the load resistance. You will use this information to calculate average load voltage, current, and power dissipation where,

$V_{P(load)} \sim V_{P(source)} - 1.4\ V$ $\quad V_{avg} = 0.637V_{P(load)}$ $\quad I_{avg} = V_{avg}/R_L$ $\quad P_D = I_{avg} \cdot V_{avg}$

$V_{P(sec)} = 1.414V_{sec(RMS)}$ and $V_{P(source)} = V_{P(sec)}$

Figure 2

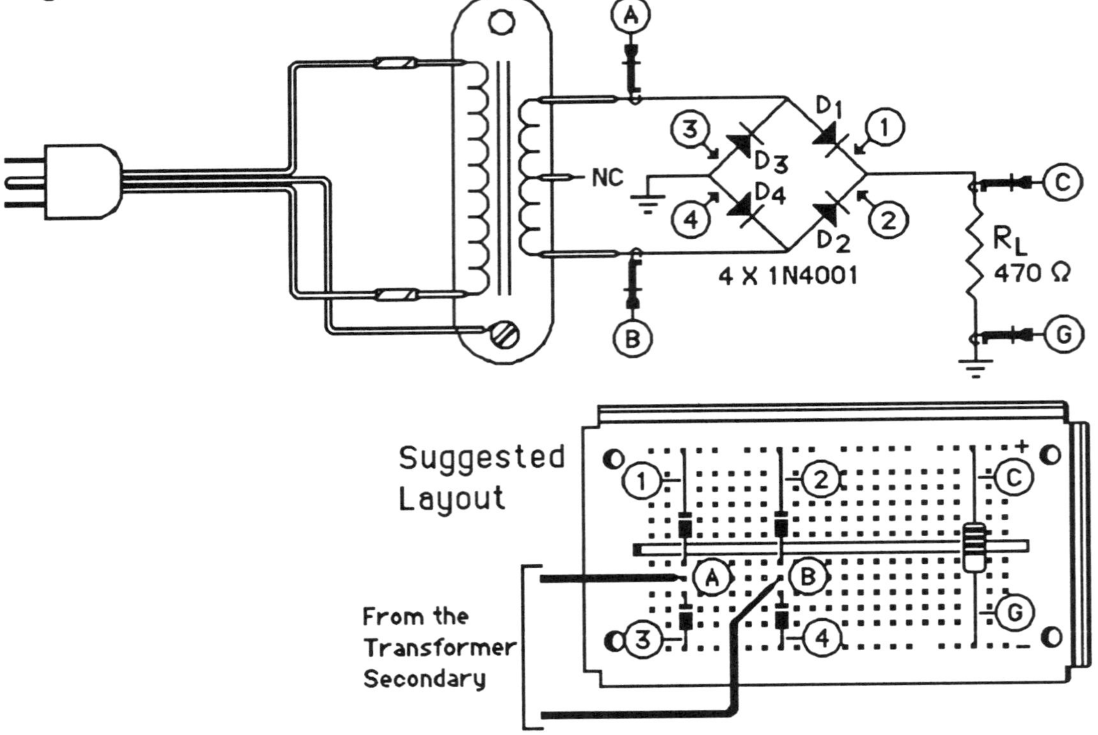

1. Assemble the bridge rectifier circuit shown in Figure 2.
2. Power-on the oscilloscope and connect a probe cable to Channel 1. Connect the Channel 1 probe to Testpoint A with ground clip to Testpoint B. This will enable you to measure the total peak-to-peak secondary voltage.
3. Plug in the power cord for the transformer. Adjust the time/division selector so approximately two complete AC cycles are shown on the screen. Set the volts/division selector and the vertical position control so the AC cycles occupy most of the screen. Measure the peak-to-peak voltage and the time per cycle and record them on the answer sheet in Part Two.
4. Important: Remove the Channel 1 probe from Testpoints A and B.
5. Use the Channel 1 probe to measure the peak value of pulsating DC across the load resistor (from Testpoint C to G).
6. On the answer sheet in Part Two, draw the waveform that you see on Screen 3 and fill in the requested data. Measure and record the peak DC voltage and the time per pulse.
7. Remove D_1 from the circuit. Use the answer sheet to describe what you see on the oscilloscope and why.
8. Place D_1 back in the circuit.

Caution: In the following steps, be sure to observe polarity when inserting the ammeter in the circuit. Always remove power while connections are being made to avoid an accidental short. Make sure all connections are secure and correct before applying power.

9. Unplug your transformer. Use your multimeter as an ammeter and place it in series with the load resistor at Testpoint C. Once all connections are safely made, plug the power cord back in. Measure the current and record the reading on the answer sheet in Part Two.

10. Unplug your transformer. Use your multimeter as an ammeter and place it in series with D_1 at the point indicated by the circled 1. Once all connections are safely made, plug the power cord back in. Measure the current and record the reading on the answer sheet in Part Two.

11. Repeat Step 10 for each diode, removing power and placing the ammeter in series with the diode at the point indicated by the circled number.

12. Unplug your transformer, turn off the oscilloscope, and answer all remaining questions for Part Two.

NAME:______________________ CLASS:__________ DATE:__________

PART ONE QUESTIONS

1. $V_{sec(RMS)} = 25.2$ V, $V_{P(sec)} =$ 35v ✓, $V_{P(source)} =$ 12v ✓

Screen 1

Ref. Line —

Ref. Line —

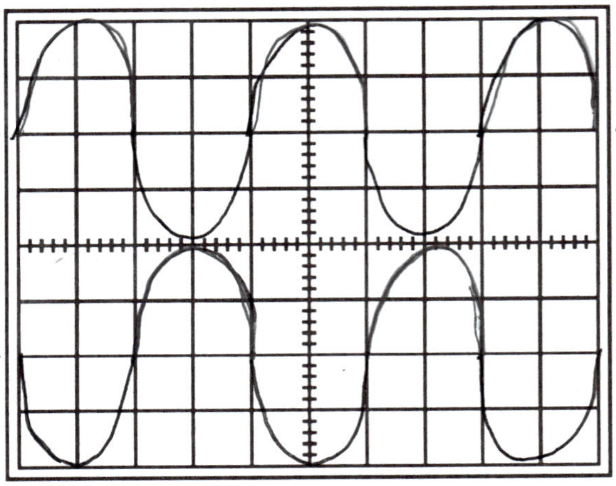

Oscilloscope Control Settings:

—Volts/Div : 10v/div ✓

Time/Div : 5ms ✓

—Volts/Div : 10v/div ✓

Screen 2

Ref. Line —

Ref. Line —

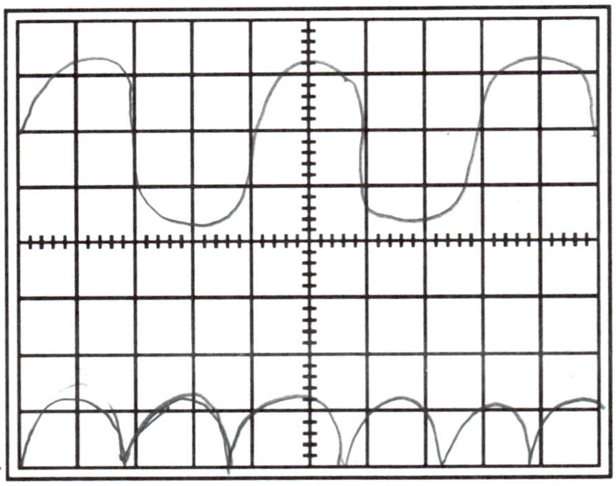

—$V_{P(source)}$ = 30v

—$V_{P(load)}$ = 15v

2. Describe what you saw when you removed D_1 from the circuit. Also, what circuit troubles would removing a diode simulate?

~~A half wave recti~~ A rectified Sine Wave ?

It would look like a half-wave rectifier - it would simulate a burnt open diode /

3. Measured value of load (current) is 21.7mA.
4. Measured current for D_1 is 10.9mA.
5. Measured current for D_2 is 10.9mA.
6. Do the sum of the currents recorded in Questions 4 and 5 equal the total measured load? Yes
7. Do your calculated values from Question 1 *exactly* match the measured values recorded next to Screen 2? If not, why not? yes
8. Using the measured peak load voltage, calculate the average DC load voltage. 12.0v
9. Using the calculated value of DC load voltage, calculate the average DC load (current). 25mA
10. Calculate the power dissipated in the load resistor. 294mW
11. Calculate the average current supplied through each diode. 12.5mA
12. Explain how this circuit can be converted into a negative full-wave rectifier. Turning the Diodes around.

PART TWO QUESTIONS

1. What is the peak-to-peak AC voltage between Testpoints A and B in Figure 2? 70v
2. What is the time per cycle (T) for the AC voltage? 17ms
3. Calculate the frequency (f) of the AC voltage where f = 1/T. 58.7Hz
4. What is the peak AC voltage of the secondary? 35v

Screen 3

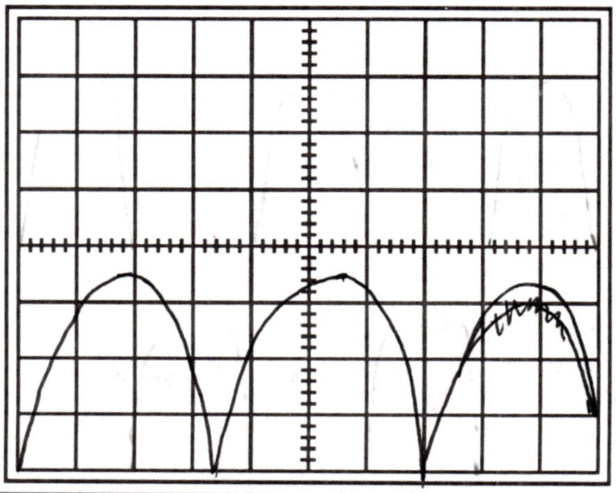

Time/Div = 5ms

T = 17ms
Time per pulse

Volts/Div = 10v/div

$V_{P(load)}$ = 35v

5. Use the time per pulse (T) recorded next to Screen 3 to calculate the DC pulse repetition rate where PRR = 1/T. 58.82 Hz

6. How does the pulse repetition rate compare to the frequency of the AC voltage?
the frequency is about the same

7. How does the peak secondary voltage of Question 4 compare with the peak DC voltage recorded next to Screen 3? Why? The same because the whole transformer is being used

8. Describe what you saw when you removed D_1 from the circuit. Also, what circuit troubles would removing a diode simulate?
Would simulate a two diode half-wave rectifier, it would be like one of the diodes was burnt open.

9. Measured value of load (current) is 49.7mA.

10. I_{D1} = 24.8mA, I_{D2} = 24.7mA, I_{D3} = 24.7mA, I_{D4} = 24.8mA

11. How does the amount of current through each diode compare to the total load (current)?
each diode has 1/2 the total current

12. Using the measured peak load voltage, calculate the average DC load voltage. 24.75V

13. Using the calculated value of DC load voltage, calculate the average DC load (current).
53mA

14. Calculate the power dissipated in the load resistor. 1.3 W

15. Calculate the average current supplied through each diode 24.25mA

16. Explain how this circuit can be converted into a negative bridge rectifier.
Turn the Diodes around

Draw the negative bridge rectifier here.

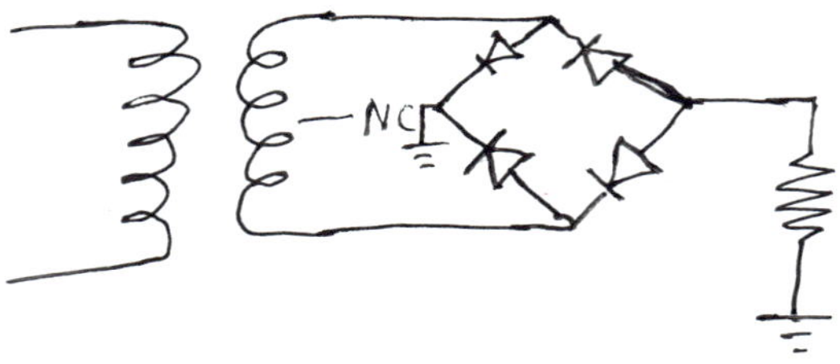

Filtering Pulsating DC

EXPERIMENT

4

PURPOSE

The purpose of this experiment is to investigate the capacitor-input filter used in many AC to DC converters. In this investigation, you will see the effects of using different capacitor values under different load conditions. Also you will see the advantage of full-wave rectification over half-wave rectification in terms of percent ripple at the load.

FORMAL OBJECTIVES

After completing this experiment, you will be able to:

- analyze capacitor-input filters to determine percent ripple voltage across the load device.
- describe the effect on percent ripple voltage that results from changes in capacitor values, changes in load, and loss of full-wave rectification.

PREPARATIONS

In preparation for this experiment, you should study Section 2-5 of *Exploring Electronic Devices*.

COMPONENTS and EQUIPMENT

1 - Dual-trace oscilloscope with test probes
1 - Protoboard
1 - Transformer with attached power cord, 25.2 V, 450 mA, center-tapped secondary
2 - 1N4001 Diodes
1 - 22 Ω, 1/4 W Resistor
1 - 220 Ω, 1/4 W Resistor
1 - 2.2 kΩ, 1/4 W Resistor
1 - 10 μF, 35 V Electrolytic Capacitor
1 - 100 μF, 50 V Electrolytic Capacitor
1 - 1,000 μF, 35 V Electrolytic Capacitor

PART ONE: THE EFFECTS OF CAPACITOR VALUE AND LOAD ON RIPPLE

In this part of the experiment, you will assemble the circuit shown in Figure 1 and use the oscilloscope to view the DC waveform across the load resistor. You will investigate the effect different values of capacitance and load have on the amount of ripple present across the load resistor. This will give you a "feel" for the reasonable value of capacitance needed to keep ripple at a minimum for a given load. Formulas that apply to this experiment are as follows:

$V_{DC} = V_{P(load)} - (V_{r(P-P)}/2)$ $\qquad$ V_r = RMS Ripple Voltage = $0.289 \cdot V_{r(P-P)}$

% Ripple = $100\% \cdot V_r/V_{DC}$

Notice the circuit of Figure 1 has a 22-Ω series resistor R_S. This resistor is not to be confused with the series resistance calculated in the many design examples in Section 2-6 of *Exploring Electronic Devices*. The purpose for this resistor is *not* to drop voltage to provide a specific output voltage. This series resistor is intended as a protection device in the event the DC output is accidentally shorted to ground. As you know, the transformer's secondary is rated for a maximum continuous average current of 450 mA. If the output becomes shorted, the combined resistance of the secondary winding and the 22-Ω resistor will insure the secondary's current remains well below 450 mA. the 22-Ω resistor is only 1/4 W which means it will be sacrificed in the event of an extended short. Therefore, the 22-Ω resistor acts as an inexpensive very-slow-blow fuse. *Take care because this resistor will be very warm, and even hot, during this experiment, especially if you short the output.*

Figure 1

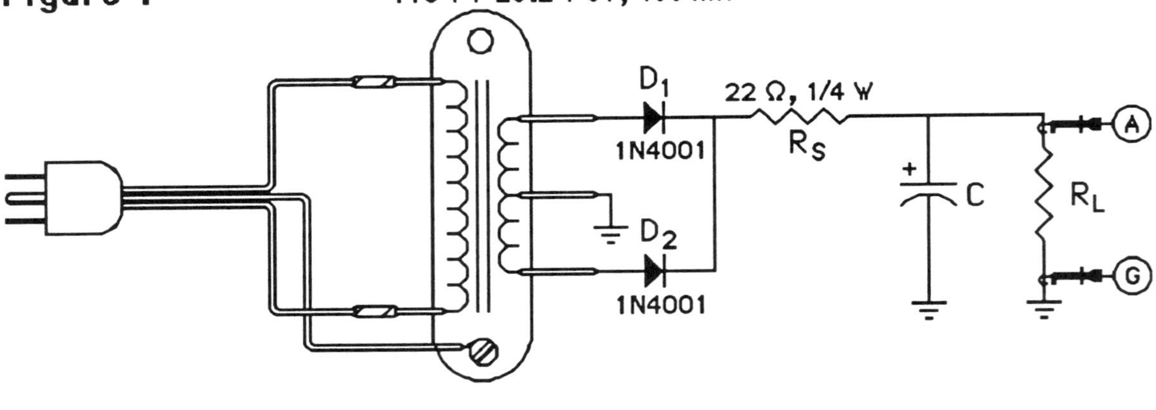

1. Assemble the circuit of Figure 1. Use a 1,000-µF capacitor and a 2.2-kΩ load resistor.

2. Connect the Channel 1 oscilloscope probe across the load resistor. Preset the oscilloscope controls as follows:

Channel 1 Input Mode Select:	DC
Trace Mode:	Channel 1 (CH1) only
Trigger Mode Select:	Auto
Trigger Source:	Channel 1
Time/Div Control:	2 ms/Div
Volts/Div Control:	5 V/Div
Horizontal Position Control:	Centered

3. Switch on the oscilloscope. Plug in the power cord for the transformer. Momentarily switch the Channel 1 Input Mode Select switch to the GND position and adjust the Vertical Position Control so the trace rests on the bottom line on the screen. This establishes the bottom line as 0 V. Switch back to the DC position. A DC level ($V_{P(load)}$) should now be represented on the CRT screen.

4. Examine Table 1 on the answer sheet for Part One. As you can see, you will measure and record $V_{P(load)}$ and $V_{r(P\text{-}P)}$. Using this data, you will calculate V_{DC}, V_r, and % Ripple. This is to be done for all combinations of capacitance and load resistance using a 10-µF, 100-µF, and a 1,000-µF capacitor and a 220-Ω and 2.2-kΩ load resistor.

   Note: When measuring the peak-to-peak ripple voltage, you may need to change the Channel 1 Input Mode Select switch to the AC position and decrease the Volts/Div control setting.

5. Unplug the transformer and switch off the oscilloscope. Answer all questions for Part One.

PART TWO: FULL-WAVE VS HALF-WAVE RECTIFICATION

In this part of the experiment, you will see what happens to load voltage and ripple when one of the diodes in the full-wave rectifier becomes open or disconnected. This part of the experiment will serve as a troubleshooting exercise and a direct comparison of full-wave and half-wave filtered AC to DC converters.

1. Remove D_1 from the circuit of Figure 1.

2. Switch on the oscilloscope. Plug in the power cord for the transformer.

3. Measure $V_{P(load)}$ and $V_{r(P\text{-}P)}$ as you did in Part One using the 100-µF capacitor and the two load resistors (one at a time).

4. Record all readings and calculations in Table 2 for Part Two.

5. Unplug the transformer and switch off the oscilloscope. Answer all questions for Part Two.

NAME: __________ **CLASS:** __________ **DATE:** __________

PART ONE QUESTIONS

Table 1

	R_L = 2.2 kΩ, 1/4 W	R_L = 220 Ω, 2 W
C = 1,000 µF	$V_{P(load)}$ = 15v $V_{r(P-P)}$ = 48mv V_{DC} = 14.976v V_r = 13.9mV % Ripple = .09%	$V_{P(load)}$ = 12v $V_{r(P-P)}$ = 290mv V_{DC} = 11.85v V_r = 84mV % Ripple = .71%
C = 100 µF	$V_{P(load)}$ = 15v $V_{r(P-P)}$ = .4v V_{DC} = 14.80v V_r = 115mv % Ripple = .78%	$V_{P(load)}$ = 12v $V_{r(P-P)}$ = 2.4v V_{DC} = 10.8v V_r = .694v % Ripple = 6.4%
C = 10 µF	$V_{P(load)}$ = 15v $V_{r(P-P)}$ = 3.25V V_{DC} = 13.4 V_r = 939mV % Ripple = 7%	$V_{P(load)}$ = 12v $V_{r(P-P)}$ = 10.8v V_{DC} = 6.6v V_r = 3.12v % Ripple = 47%

1. Explain the purpose for the 22-Ω series resistor in the circuit of Figure 1.
   to provide a specific output voltage

2. Which provides the greatest load, (a) the 220-Ω load resistor, or (b) the 2.2-kΩ resistor ?
   the 220 Ω load resistor

3. Which load resistor caused the greatest amount of ripple voltage? the 220 Ω

4. Which capacitor provided the best filtering (the least ripple)? 1000µF

5. Why was $V_{P(load)}$ for the 220-Ω resistor slightly less than $V_{P(load)}$ for the 2.2-kΩ resistor?
   Because the was more of a voltage drop across the 22 Ω resistor when the 220 Ω is in the circuit (voltage divider rule)

6. To minimize ripple, the greater the load, the
(a) larger the filter capacitor. (b) smaller the filter capacitor.
"larger" means more capacitance "smaller" means less capacitance

PART TWO QUESTIONS

Table 2 (D_1 removed)

	R_L = 2.2 kΩ, 1/4 W	R_L = 220 Ω, 2 W
C = 100 µF	$V_{P(load)}$ = 30V	$V_{P(load)}$ = 17V
	$V_{r(P-P)}$ = 1.75V ?	$V_{r(P-P)}$ = 10V
	V_{DC} = 29.125V	V_{DC} = 12V
	V_r = 506mV	V_r = 2.89V
	% Ripple = 1.7%	% Ripple = 24%

1. Describe how your findings in Table 2 compare to those of Table 1.
The measurements in table 2 were much Higher - the ripple voltages were much more prominent

2. Explain why the percent ripple is greater in Table 2 than in Table 1.
Because in a half wave rectifier the waves are spread out more, therefore the slope can drop for a longer time period

3. Which is more desirable for a filtered AC to DC converter?
(a) half-wave rectifier (b) full-wave rectifier

4. Assuming the same value filter capacitor and load, which of these generates more ripple voltage? (a) half-wave rectifier (b) full-wave rectifier

5. Describe the effect a leaky capacitor would have on ripple voltage. A leaky capacitor is a capacitor that has an internal conductive path that acts as a load to discharge the capacitor.
The filtering capability would not be as good as an fully operational capacitor

NOTES AND CALCULATIONS

Voltage Regulation - The Zener Diode

EXPERIMENT 5

PURPOSE

The purpose of this experiment is to demonstrate the use of the zener diode as a voltage regulator and to investigate and prove circuit theory.

FORMAL OBJECTIVES

After completing this experiment, you will be able to:

- analyze or evaluate zener-diode voltage regulators.
- determine the percent regulation for a zener-diode regulator from measured values of full-load and no-load zener voltage.
- determine the minimum value of load resistance for a given zener regulator circuit.

PREPARATIONS

In preparation for this experiment, you should study Sections 3-1 and 3-2 of *Exploring Electronic Devices.*

COMPONENTS and EQUIPMENT

1 - Analog or digital multimeter
1 - Protoboard
1 - Transformer with attached power cord, 25.2 V, 450 mA, center-tapped secondary
2 - 1N4001 Diodes
1 - 1,000 μF, 35 V Capacitor
1 - 1N4742, 12 V, 1 W Zener diode
1 - 120 Ω, 1 W Resistor
1 - 220 Ω, 2 W Resistor
1 - 330 Ω, 1 W Resistor

DISCUSSION

In the following steps, you will assemble and analyze the AC to DC converter and zener-diode voltage regulator circuit of Figure 1. You will compare calculated values of voltage and current with actual measured values in order to verify circuit theory. Also, you will discover how well a zener regulator performs under the extreme conditions of no load and full load.

Caution: The zener diode, the series resistor, and the load resistor will be hot to the touch during the experiment.

Recall from your reading that the zener diode conducts that portion of the total current (I_{RS}) that is not part of the load (I_L) where $I_Z = I_{RS} - I_L$. The total current through R_S remains nearly constant for any load condition from full load to no load.

$V_{DC(in)} = V_C = (1.414 \bullet V_{sec}/2) - 0.7\ V$ $\qquad V_{RS} = V_{DC(in)} - V_Z$ $\qquad I_T = I_{RS} = V_{RS}/R_S$

$\%\ Regulation = 100\% \bullet (V_{NL} - V_{FL})/V_{FL}$

Figure 1

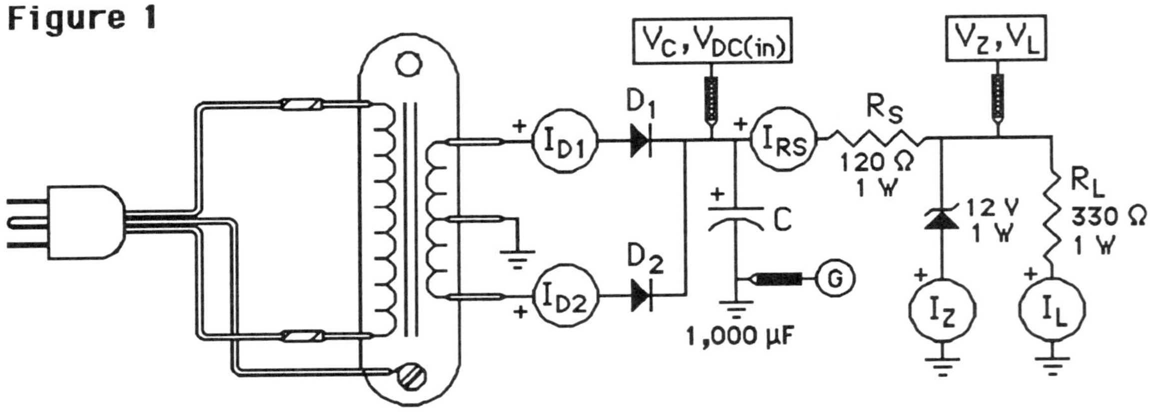

PROCEDURE

1. Construct the circuit shown in Figure 1.

2. Before you plug in the transformer's power cord, take time to perform a theoretical analysis on your circuit. Place all calculated values in the first column of Table 1 on the answer sheet. Use the open-page area labeled "Notes and Calculations" to show your work.

3. Now you are ready to perform an experimental analysis of your circuit and compare your results with the calculated, or theoretical, values. Use your multimeter as a voltmeter to measure the voltages listed in Table 1. Record your readings in Table 1.

4. Use your multimeter as an ammeter to measure all currents listed in Table 1. Record your readings. The ammeter symbols shown in Figure 1 indicate the proper placement for your meter for each reading. *See caution on next page.*

Caution: Be sure to observe proper polarity when placing the ammeter in the circuit. Also, unplug the transformer's power cord while connecting the meter in the circuit. This will help avoid damage to the diodes and the transformer due to accidental shorts.

5. Use the measured values of voltage and/or current to calculate the power dissipated by the series resistor, the zener diode, and the load resistor. Record your answers on the answer sheet.

6. We will consider the load (current) through the 330-Ω resistor to be full load for the zener-diode regulator. V_Z, measured and recorded in Table 1, is the full-load voltage (V_{FL}). Remove the 330-Ω load resistor and measure the zener voltage with no load (V_{NL}). Record your reading on the answer sheet.

7. Calculate the percent regulation and record your answer on the answer sheet.

8. Replace the 330-Ω load resistor with a 220-Ω resistor. Measure V_Z, I_Z, and I_L. Record your readings on the answer sheet.

9. Unplug the power cord of your transformer. Answer all remaining questions on the answer sheet.

NOTES AND CALCULATIONS

NAME: Doug + Jesse **CLASS:**________ **DATE:**________

QUESTIONS

Table 1

	Calculated Values	Measured Values
$V_{DC(in)}$	17.12v	16.4v
V_{RS}	5.12v	4.7v
V_Z, V_L	12 V	11.7v
I_{RS}	43ma	38.9mA
I_L	36ma	36.2mA
I_Z	7ma	3.7mA
I_{D1}	54mA	48.3mA
I_{D2}	54mA	48.3mA

1. P_{RS} = 184.32 mW, P_{RL} = 432mW, P_Z = 84 mW
2. V_{NL} = 12.3V
3. % Regulation = 2.5%
4. What is the difference in zener voltage from full load to no load? .3 volts
5. What would the full-load and no-load output voltage be if the zener were not in the circuit? (Determine V_L with and without the 330-Ω load resistor.)
   $V_{L(\text{with 330-}\Omega\text{ resistor})}$ = 12.5v, $V_{L(\text{no load resistor})}$ = 17v
6. Determine the percent regulation using the values calculated in Question 5.
   % Regulation = 36%
7. With a 220-Ω load resistor:
   V_Z = 10.95v, I_Z = .00mA, I_L = 94.2
8. Explain the reason for the measured values of Question 7.
   No current because 10.95V is not enough to turn on the zener
   The voltage is lower because of the lower resistance - same for increase current.
9. Calculate the approximate value of load resistance below which the zener diode ceases to conduct and there is no longer any regulation. $R_{L(\text{minimum})}$ = 240 Ω

Voltage Multiplication

EXPERIMENT 6

PURPOSE

The purpose of this experiment is to familiarize you with basic half-wave and full-wave voltage multiplier circuits.

FORMAL OBJECTIVES

After completing this experiment, you will be able to:

- construct and analyze basic half-wave voltage multiplier circuits.
- construct and analyze a basic full-wave voltage doubler.

PREPARATIONS

In preparation for this experiment, you should study Section 3-3 of *Exploring Electronic Devices.*

COMPONENTS and EQUIPMENT

1 - Analog or digital multimeter
1 - Protoboard
1 - Transformer with attached power cord, 25.2 V, 450 mA, center-tapped secondary
3 - 1N4001 Diodes
3 - 100 µF, 50 V Capacitors
1 - 22 kΩ, 1/4 W Resistor

PART ONE: HALF-WAVE VOLTAGE MULTIPLIERS

In this part of the experiment, you will investigate the half-wave voltage multiplier. You will assemble and test the circuit of Figure 1. Voltage multipliers such as this are used to obtain DC voltages that are much greater than the peak voltage of the AC source. The current supply capability of these circuits decreases as the number of multiplier sections increases. As a result, these circuits are usually intended for high-voltage, low-current loads. In this experiment, we deliberately use only half of the secondary winding to keep the multiplied voltage at a safe level. Normally, the entire secondary would be used.

Caution: Do not use the entire secondary in this experiment. Doing so would be dangerous to your health and will probably destroy your capacitors.

Let's take just a moment to review this circuit's theory of operation. When power is first applied, the first voltage alternation in which TP-G is negative and TP-2 is positive will forward bias D_1 and charge C_1 to the peak AC value where $V_{C1} = V_P$. On the very next alternation, where TP-G is positive and TP-2 is negative, D_1 is reverse biased and D_2 becomes forward biased. The positive peak voltage at TP-G is added (series aiding) to the charge on C_1 and is used to charge C_2 via D_2. Thus, $V_{C2} = 2V_P = 2V_{C1}$. On the very next alternation, where TP-G is once again negative and TP-2 is once again positive, D_2 is reverse biased and D_3 becomes forward biased. D_1 will become forward biased only for a short time at the peak of the alternation to replace any charge that has been lost from C_1. Now, because TP-2 is heading toward a positive peak, it is series aiding with

Figure 1

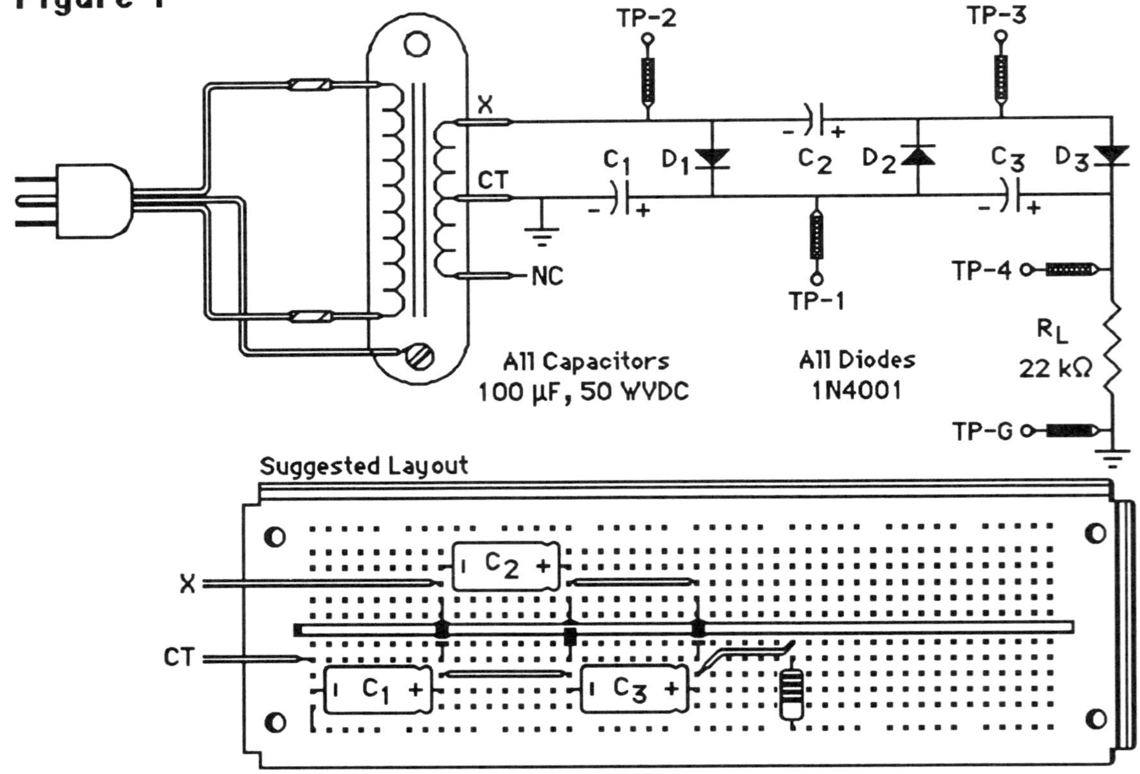

the charge on C_2. The combined potential is used to charge C_3 to $2V_P$ where $V_{C3} = V_P + V_{C2} - V_{C1}$ because V_{C1} is holding the negative lead of C_3 at a positive potential equal to V_P. The total output voltage from TP-4 to TP-G is equal to $3V_P = V_{C1} + V_{C3} = V_{RL}$. From this, you can see that V_{C1} is always equal to V_P and the charge on all other capacitors (V_{C2}, V_{C3} . . .) is equal to $2V_P$. Thus, the voltage rating of each capacitor should be greater than $2V_P$.

1. Assemble the circuit of Figure 1.

   Important: Have someone check your circuit to make sure it is wired correctly before continuing. Capacitors will be destroyed if they are placed incorrectly.

2. Plug in the power cord for the transformer.

3. Use your multimeter to measure the voltage across each capacitor. Record your findings on the answer sheet in Part One.

4. Use your multimeter to measure the voltage between all testpoints and ground (TP-G). Record your findings on the answer sheet in Part One.

5. Measure the voltage across the load resistor. Record the reading on the answer sheet for Part One and explain which capacitors provide this load voltage.

6. Unplug the transformer. Answer all remaining questions for Part One.

   Optional: Use the oscilloscope to view the DC voltage across the load resistor. Change the load resistor to different values and observe any change in the amount of ripple voltage. Take care in doing so because any resistor values less than 22 kΩ will become hot.

PART TWO: FULL-WAVE VOLTAGE MULTIPLIERS

In this part of the experiment, you will assemble and test a basic full-wave voltage doubler circuit. As shown in Figure 2, the capacitors are charged of opposite polarity via the diodes. The load is placed across both capacitors to obtain a DC voltage equal to $2V_P$ where C_1 and C_2 are series aiding. Once again, we use only half of the secondary to keep the voltage at a safe level for you and your capacitors. *Do not use the full secondary!*

Figure 2

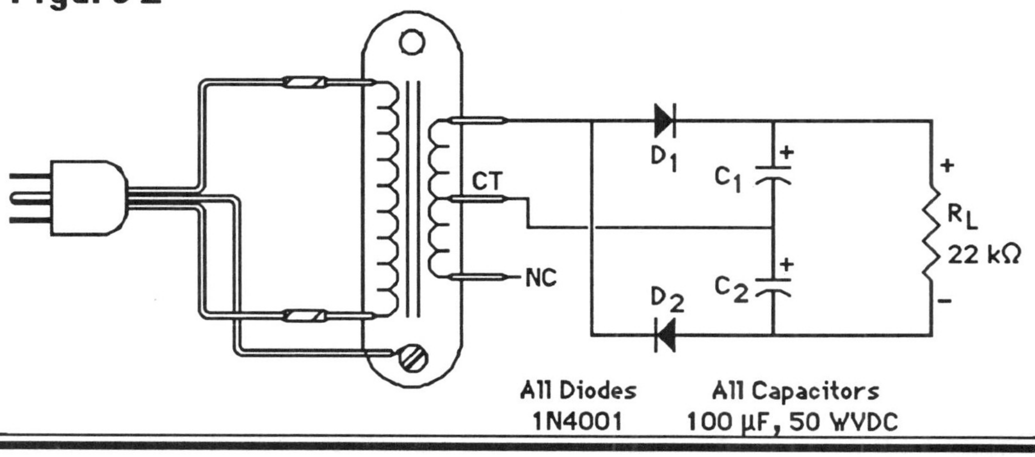

1. Assemble the full-wave voltage doubler shown in Figure 2.

   Important: Have someone check your circuit to make sure it is wired correctly before continuing. Capacitors will be destroyed if they are placed incorrectly.

2. Plug in the power cord for the transformer.

3. Use your multimeter to measure the voltage across each capacitor. Record your findings on the answer sheet in Part One.

4. Measure the voltage across the load resistor. Record the reading on the answer sheet for Part One.

5. Unplug the transformer. Answer all remaining questions for Part Two.

NOTES AND CALCULATIONS

NAME:______________________ CLASS:__________ DATE:__________

PART ONE QUESTIONS

1. V_{C1} = 16.2v, V_{C2} = 32.3v (RMS), V_{C3} = 32.1v,
2. $V_{TP\text{-}1}$ = 16.1v, $V_{TP\text{-}2}$ = 0v, $V_{TP\text{-}3}$ = 32.3v,
   $V_{TP\text{-}4}$ = 48.1 (Vp-p)
3. V_{RL} = 48.1v
4. Which capacitors provide the load voltage? None
5. Is the circuit of Figure 1 a voltage (a) doubler, (b) tripler, or (c) quadrupler ?
6. Draw a half-wave voltage quadrupler circuit in this space. Clearly mark the terminal at which $V_{DC} = 4V_P$.

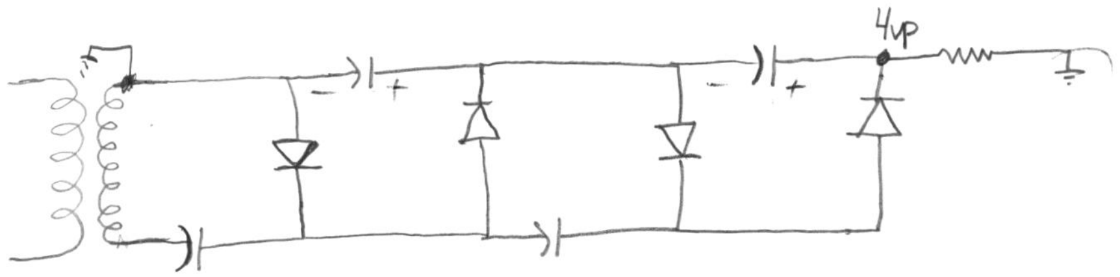

PART TWO QUESTIONS

1. V_{C1} = 16.3v, V_{C2} = 16.3v
2. V_{RL} = 32.7
3. In Figure 2, which diode charges C_1? D_1
4. If the load resistor is removed, will D_1 or D_2 continue to conduct with each alternation of the secondary voltage? Explain.
   Both diodes continue to conduct and charge the capacitors Because current flows through the D_1 and C_1 to ground. D_2 conducts because current flows through D_1 + C_1 + C_2 + D_2 + D_1 + C_1 to ground

5. Is the capacitor voltage exactly equal to the peak secondary voltage? Explain.
The capacitor voltage was the same because the capacitor charges to the peak of the secondary
Is this significant in a high-voltage multiplier circuit? yes

6. What would you do to this circuit to further reduce DC ripple voltage across the load resistor?
Add a bigger capacitor or increase the frequency

NOTES AND CALCULATIONS

Voltage Limiter Circuits

EXPERIMENT

7

PURPOSE

The purpose of this experiment is to give you the opportunity to explore a variety of voltage-limiter (clipper) circuits.

FORMAL OBJECTIVES

After completing this experiment, you will be able to:

- determine if a limiter circuit is a positive, negative, or combination voltage limiter from a schematic of the circuit.
- determine if a voltage limiter circuit is symmetrical or asymmetrical from a schematic of the circuit.
- predict the shape and degree of limiting of a voltage-limited waveform from a schematic of the circuit.

PREPARATIONS

In preparation for this experiment, you should study Section 3-4 of *Exploring Electronic Devices*.

COMPONENTS and EQUIPMENT

1 - Analog or digital multimeter
1 - Protoboard
1 - Transformer with attached power cord, 25.2 V, 450 mA, center-tapped secondary
1 - Variable DC power supply
2 - 1 kΩ, 1/4 W Resistors
2 - 1N914 Switching diodes
1 - 1N4742, 12 V, 1 W Zener diode
1 - 1N4733, 5.1 V, 1 W Zener diode

DISCUSSION

In the following steps, you will assemble six different voltage limiter circuits and use the oscilloscope to observe the limited output voltage. Recall that the voltage limiter diode(s) remain in a nonconducting state until the applied voltage exceeds a voltage threshold. Once the voltage threshold is exceeded, the diode(s) conducts and hold(s) the output voltage at a constant level. In two of the limiter circuits a DC supply is used to back bias the diode in order to raise the threshold voltage.

Also, recall that symmetrical limiting is the type of limiting in which the positive and negative alternations of the AC signal are limited to the same voltage (of opposite polarity). In other words, the limited positive alternation is a mirror image of the limited negative alternation and vice versa. Conversely, asymmetrical limiting is the type of limiting in which the two alternations are limited unequally or only one alternation is limited.

PROCEDURE

1. For each of the following limiter circuits, Channel 1 of the oscilloscope is connected to Testpoints A and G while Channel 2 is connected to Testpoints B and G. The following oscilloscope presets are recommended.

Channel 1 Input Mode Select:	DC
Trace Mode:	Alternate
Trigger Mode Select:	Auto
Trigger Source:	Channel 1
Time/Div Control:	5 ms/Div
Channel 1 Volts/Div Control:	10 V/Div
Channel 2 Volts/Div Control:	5 V/Div (Change as needed)
Horizontal Position Control:	Centered

   Use the GND position of the Channel-1 and Channel-2 Input Mode Select switches and the Vertical Position controls to set the reference line for each channel at the center horizontal line of the upper and lower half of the screen (see Screen 1). Set the Mode Select switch back to the DC position for measurements.

2. Assemble each circuit in Figures 1 through 6 one at a time. Predict the positive and negative limiting levels for each circuit before using the oscilloscope to observe the output waveform.

   Caution: Always unplug your transformer before attempting to assemble a circuit. This will prevent damage to the transformer due to accidental shorts.

   In each case, have someone check your circuit to make sure it is correct before applying power.

3. Write your predictions on the answer sheet. Then, apply power to the circuit and observe the AC signal (CH-1) and the limited signal (CH-2) using the oscilloscope. Draw the waveforms on the screen that corresponds to the circuit. Fill in the positive and negative peak values next to the screen on the answer sheet.

4. Answer any questions on the answer sheet that pertain to the circuit you are testing before continuing to the next circuit.

5. Assemble the circuit of Figure 1. Predict the limiting levels. Observe the AC and limited waveforms with the oscilloscope. Record your observations on and next to Screen 1 on the answer sheet.

Figure 1

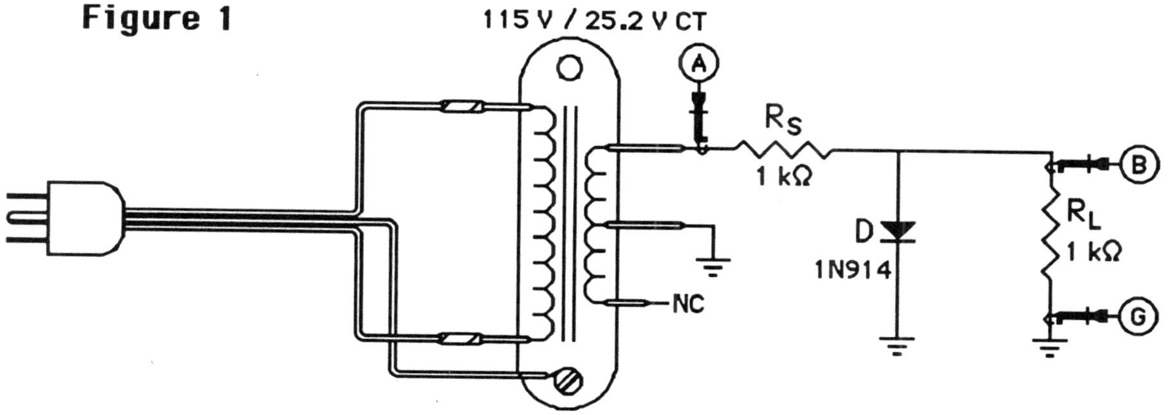

6. Assemble the circuit of Figure 2. Predict the limiting levels. Observe the AC and limited waveforms with the oscilloscope. Record your observations on and next to Screen 2 on the answer sheet.

Figure 2

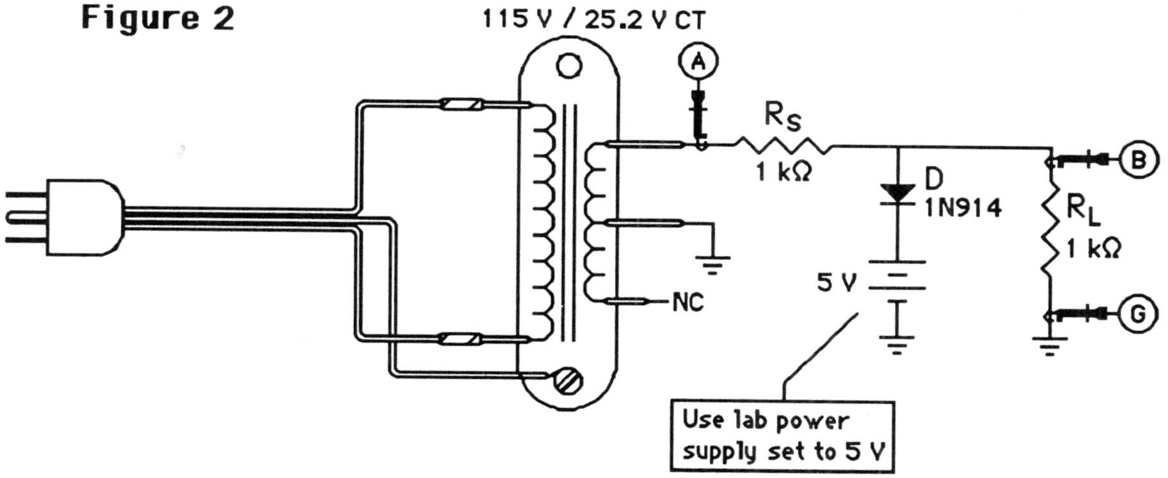

7. Assemble the circuit of Figure 3. Set the variable DC supply to -8 V. Predict the limiting levels. Observe the AC and limited waveforms with the oscilloscope. Record your observations on and next to Screen 3 on the answer sheet.

8. Slowly vary the DC voltage from 0 V to -10 V and observe the limited waveform on Channel 2. Describe what you see on the answer sheet.

Figure 3

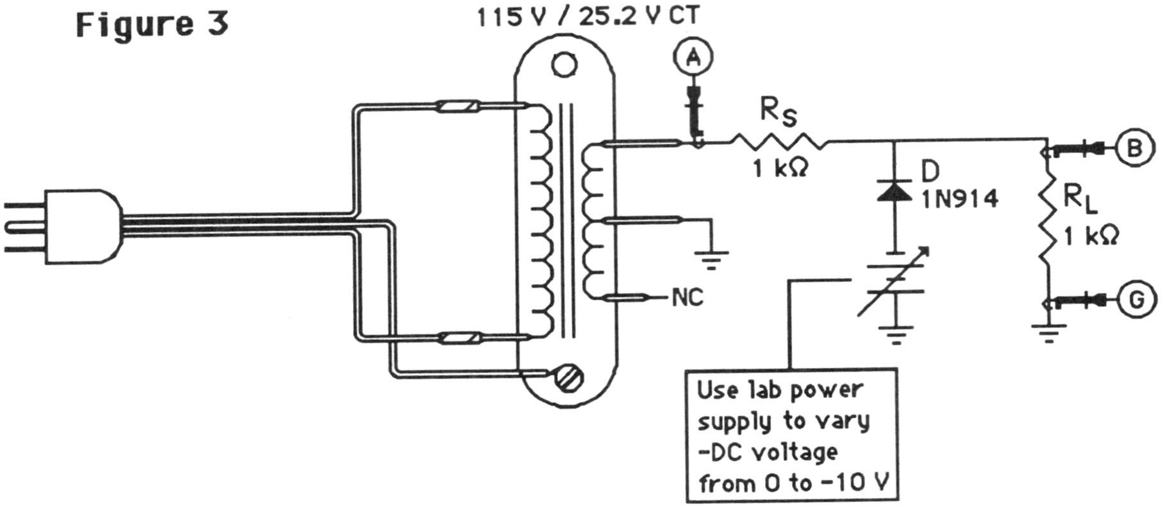

9. Assemble the circuit of Figure 4. Predict the limiting levels. Observe the AC and limited waveforms with the oscilloscope. Record your observations on and next to Screen 4 on the answer sheet.

Figure 4

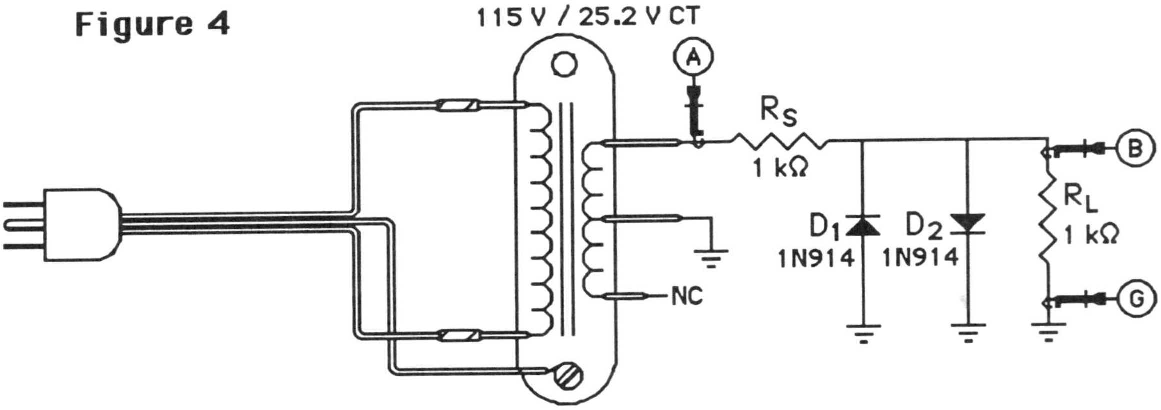

Figure 5

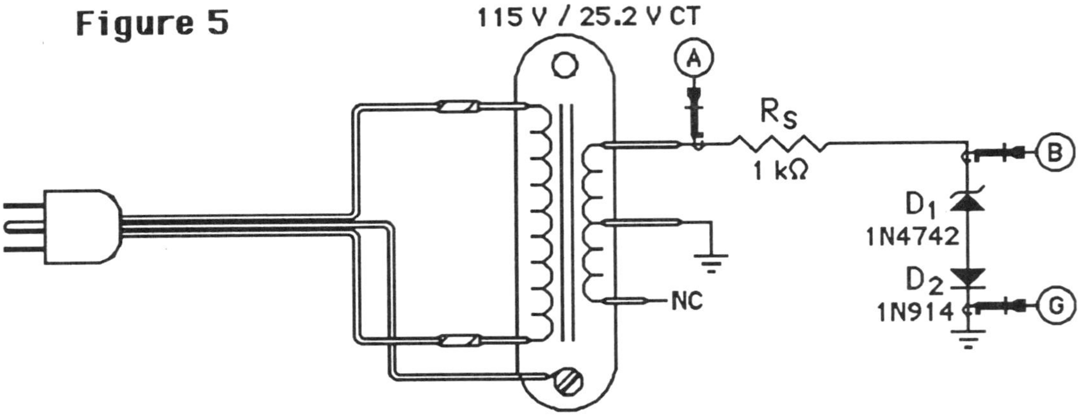

10. Assemble the circuit of Figure 5. Predict the limiting levels. Observe the AC and limited waveforms with the oscilloscope. Record your observations on and next to Screen 5 on the answer sheet.

11. Assemble the circuit of Figure 6. Predict the limiting levels. Observe the AC and limited waveforms with the oscilloscope. Record your observations on and next to Screen 6 on the answer sheet.

Figure 6

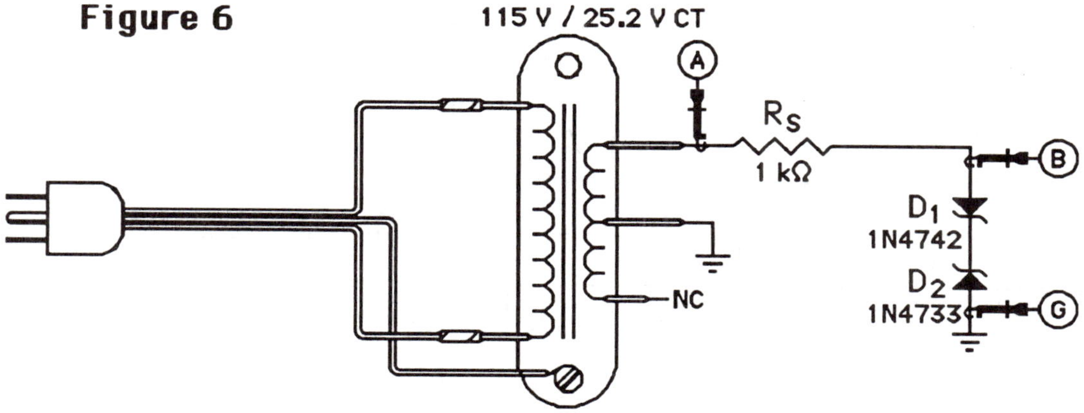

12. Unplug the power cord for your transformer. Answer any remaining questions on the answer sheet.

NAME:______________________ CLASS:__________ DATE:__________

QUESTIONS

Screen 1

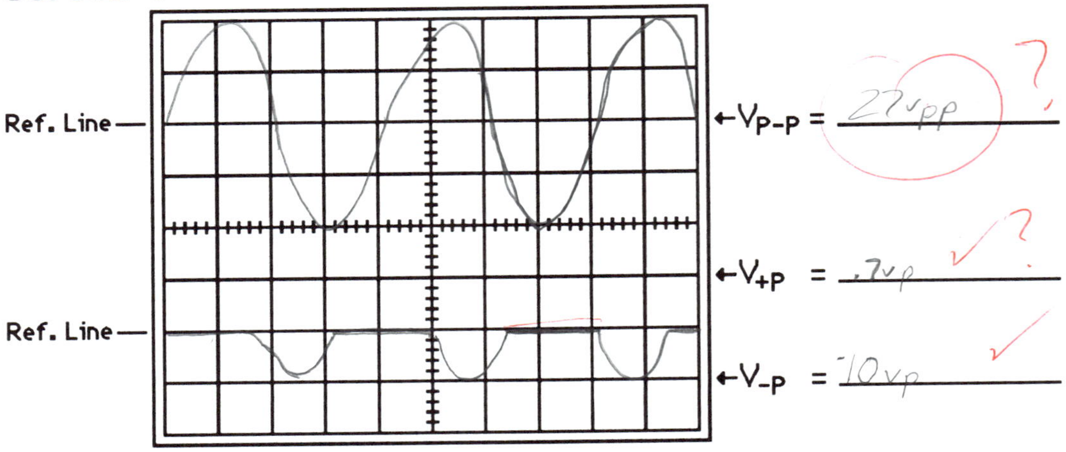

1. Predicted voltage levels for Figure 1: V_{+P} = .7v, V_{-P} = 13.5v

2. The circuit of Figure 1 is a
   (a) positive limiter. (b) negative limiter. (c) combination limiter.

3. The circuit of Figure 1 is a(n)
   (a) symmetrical limiter. (b) asymmetrical limiter.

4. During which alternation of the applied AC does the diode in Figure 1 conduct?
   (a) positive (b) negative

Screen 2

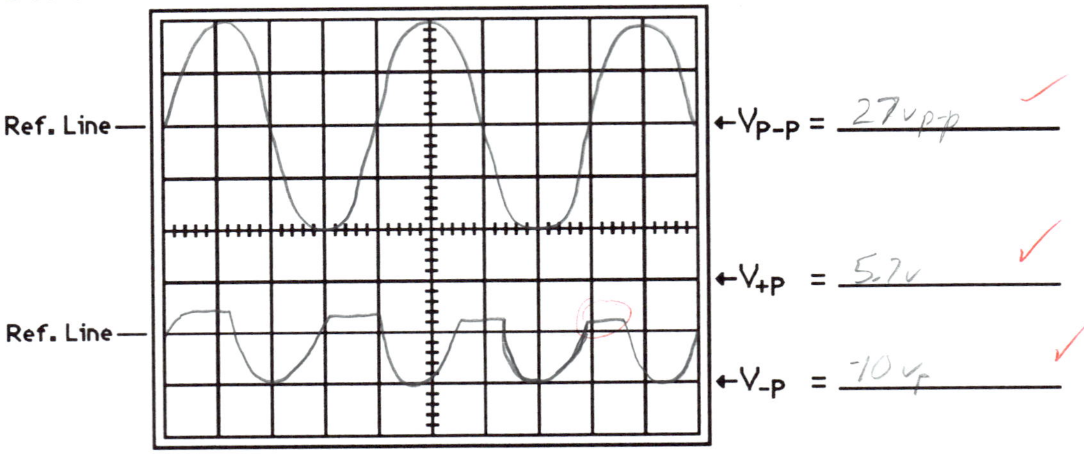

5. Predicted voltage levels for Figure 2: V_{+P} = 5.7v, V_{-P} = 13.5v

6. The circuit of Figure 2 is a
(a) positive limiter. (b) negative limiter. (c) combination limiter.

7. The circuit of Figure 2 is a(n)
(a) symmetrical limiter. (b) asymmetrical limiter.

8. During which alternation of the applied AC does the diode in Figure 2 conduct?
(a) positive (b) negative

Screen 3

Ref. Line —

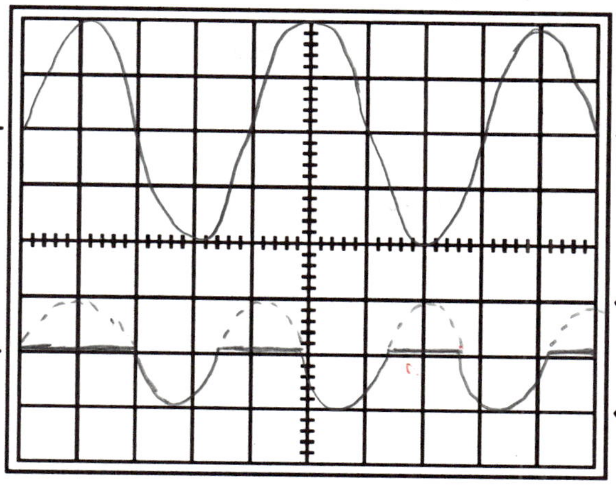

Ref. Line —

← V_{P-P} = 27vpp

← V_{+P} = .7v - 10v

← V_{-P} = -10v

inverted

9. Predicted voltage levels for Figure 3: V_{+P} = .7-10v, V_{-P} = 13.5v

10. The circuit of Figure 3 is a
(a) positive limiter. (b) negative limiter. (c) combination limiter.

11. The circuit of Figure 3 is a(n)
(a) symmetrical limiter. (b) asymmetrical limiter.

12. During which alternation of the applied AC does the diode in Figure 3 conduct?
(a) positive (b) negative

13. Describe what you observed when you varied the DC voltage in the circuit of Figure 3.
When the DC Power supply increased in Voltage the limiter decreased

Screen 4

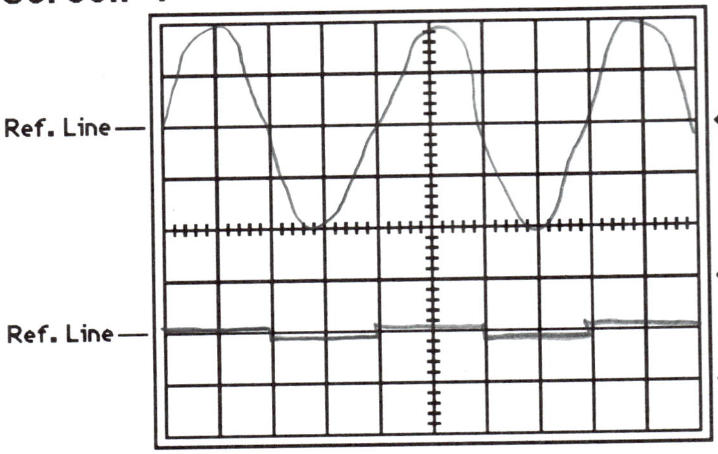

$\leftarrow V_{P-P}$ = 22vpp

$\leftarrow V_{+P}$ = .7v

$\leftarrow V_{-P}$ = -.7v

14. Predicted voltage levels for Figure 4: V_{+P} = +.7v, V_{-P} = -.7v

15. The circuit of Figure 4 is a
    (a) positive limiter. (b) negative limiter. (c) combination limiter.

16. The circuit of Figure 4 is a(n)
    (a) symmetrical limiter. (b) asymmetrical limiter.

17. During which alternation of the applied AC does D_1 in Figure 4 conduct?
    (a) positive (b) negative

Screen 5

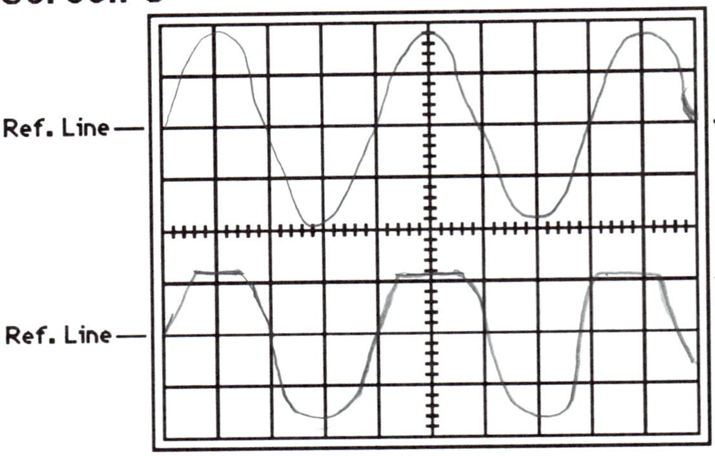

$\leftarrow V_{P-P}$ = 27vpp

$\leftarrow V_{+P}$ = 12vpp

$\leftarrow V_{-P}$ = 13vpp

18. Predicted voltage levels for Figure 5: V_{+P} = 12v, V_{-P} = 13.5vpp

19. The circuit of Figure 5 is a
    (a) positive limiter. (b) negative limiter. (c) combination limiter.

20. The circuit of Figure 5 is a(n)
(a) symmetrical limiter. (b) asymmetrical limiter.

21. During which alternation of the applied AC does D_1 act as the limiter in Figure 5?
(a) positive (b) negative

Screen 6

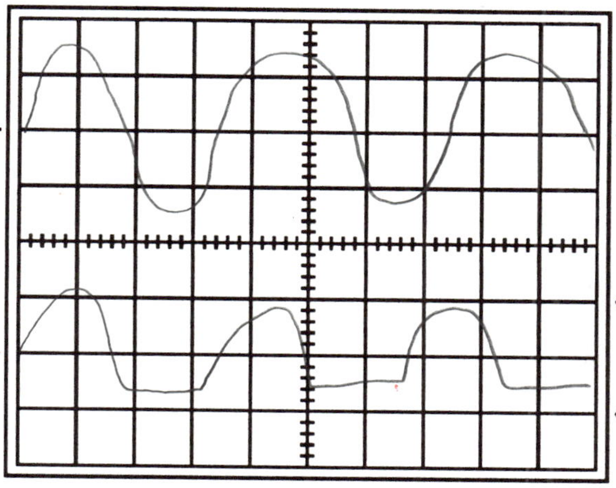

Ref. Line— ←V_{P-P} = 27v

←V_{+P} = 13.5v

Ref. Line— ←V_{-P} = 5.7v

22. Predicted voltage levels for Figure 6: V_{+P} = 13.5v, V_{-P} = 5.7v

diodes Backwards

23. The circuit of Figure 6 is a
(a) positive limiter. (b) negative limiter. (c) combination limiter.

24. The circuit of Figure 6 is a(n)
(a) symmetrical limiter. (b) asymmetrical limiter.

25. During which alternation of the applied AC does D_1 act as the limiter in Figure 6?
(a) positive (b) negative

26. Did the 0-V reference line for any of the output waveforms shift up or down on the screen from the preset center line position in the lower half of the screen? no

NOTES AND CALCULATIONS

Voltage Clamper Circuits

EXPERIMENT 8

PURPOSE

The purpose of this experiment is to familiarize you with the concept of voltage, or signal, clamping.

FORMAL OBJECTIVES

After completing this experiment, you will be able to:

- recognize basic voltage clamper circuits.
- identify positive and negative diode clamper circuits.
- predict the amount of clamping from schematic component values and voltages.

PREPARATIONS

In preparation for this experiment, you should study Section 3-4 of *Exploring Electronic Devices*.

COMPONENTS and EQUIPMENT

1 - Dual-trace oscilloscope
1 - Protoboard
1 - 12 V DC source
1 - 1 μF, 25 WVDC Capacitor
1 - Multimeter
1 - Function generator
2 - 10 kΩ Resistors
1 - 1N914 Switching diode

PART ONE: VOLTAGE-DIVIDER CLAMPERS

In this part of the experiment, you will assemble the voltage-divider clamper shown in Figure 1. As you know from your reading, a clamper is a circuit that creates a DC reference for the AC signal. At the output of the clamper, the AC signal no longer has a 0 V reference. Normally, the AC signal is untouched as far as waveform is concerned. The AC signal is merely shifted in the positive or negative direction. The signal may be described as having a positive or negative bias. Both alternations of the AC signal may or may not be totally in the positive range or the negative range. It all depends on the amount of clamping or bias as compared to the amplitude of the AC signal. The clamped waveform is also described as an AC signal that is superimposed on a DC reference. Later, in your study of transistors, you will see how voltage divider clamping is used to bias an AC signal in order to drive a transistor amplifier.

Figure 1

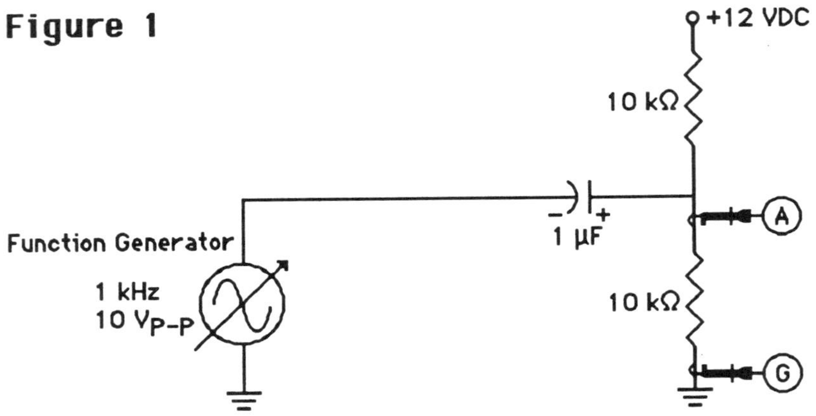

1. Assemble the circuit shown in Figure 1.
2. Apply 12 VDC to the voltage divider as shown.
3. Apply power to the function generator. With the function generator connected to the circuit, as shown, temporarily connect the Channel 1 oscilloscope probe to the generator output and adjust the generator's controls to obtain a 10-V_{P-P}, 1-kHz, sine-wave signal. The Time/Div control on the oscilloscope should be set at 0.5 ms/Div and the Volts/Div control should be set at 2 Volts/Div.
4. Move the Channel 1 probe to Testpoints A and G.
5. Temporarily set the Input Mode Select switch on the oscilloscope to the GND position and use the Vertical Position control to set the trace on the bottom line as the 0-V reference line. See Screen 1 on the answer sheet.
6. Switch the Input Mode Select switch on the oscilloscope to the DC position. Draw what you see on Screen 1.
7. Remove all power and answer all questions for Part One on the answer sheet.

PART TWO: BASIC DIODE CLAMPER CIRCUITS

In this part of the experiment, you will assemble and test a positive and a negative diode clamper circuit. These are the clampers you read about in Section 3-4 of *Exploring Electronic Devices*. In these type clampers, the clamping action is automatic and varies with the amplitude of the applied AC signal where the DC clamping voltage is equal to approximately V_P minus one diode drop. This type clamping action is seen at the base terminal of Class C transistor amplifiers. You will study this later.

Figure 2

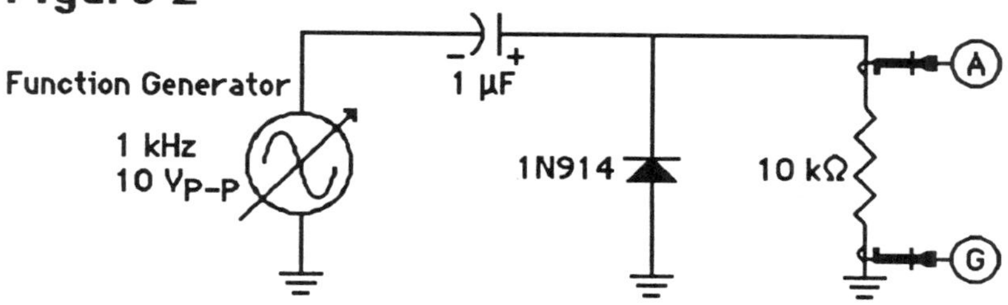

1. Assemble the circuit of Figure 2. As you can see, no DC supply is needed.
2. Apply power to the function generator. With the function generator connected to the circuit, as shown, temporarily connect the Channel 1 oscilloscope probe to the generator output and adjust the generator's controls to obtain a 10 $V_{P\text{-}P}$, 1 kHz, sine wave signal. The Time/Div control on the oscilloscope should be set at 0.5 ms/Div and the Volts/Div control should be set at 2 Volts/Div.
3. Move the Channel 1 probe to Testpoints A and G.
5. Temporarily set the Input Mode Select switch on the oscilloscope to the GND position and use the Vertical Position control to set the trace 1 cm up from the bottom line as the 0-V reference line. See Screen 2 on the answer sheet.
6. Switch the Input Mode Select switch on the oscilloscope to the DC position. Draw what you see on Screen 2. Record the positive and negative peak voltage values next to the screen in the space provided.
7. Use your multimeter to measure the DC charge voltage directly across the capacitor. Record your reading on the answer sheet.
8. Turn off the generator. Answer all questions pertaining to Figure 2.

Figure 3

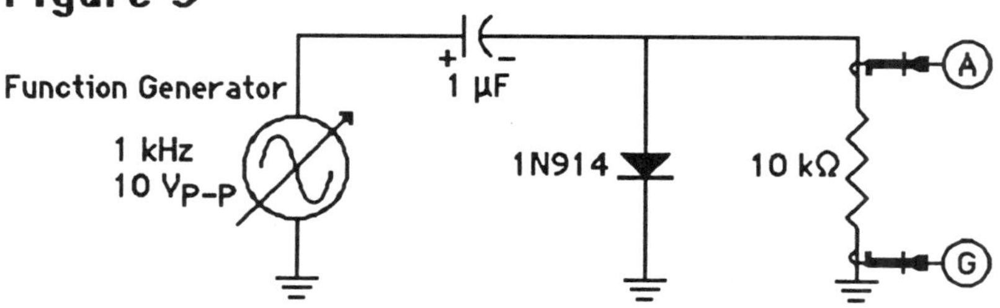

9. Assemble the circuit of Figure 3. (Reverse your capacitor and diode.)

10. Apply power to the function generator. With the function generator connected to the circuit, as shown, temporarily connect the Channel 1 oscilloscope probe to the generator output and adjust the generator's controls to obtain a 10 V_{P-P}, 1 kHz, sine wave signal. The Time/Div control on the oscilloscope should be set at 0.5 ms/Div and the Volts/Div control should be set at 2 Volts/Div.

11. Move the Channel 1 probe to Testpoints A and G.

12. Temporarily set the Input Mode Select switch on the oscilloscope to the GND position and use the Vertical Position control to set the trace 1 cm down from the top line as the 0-V reference line. See Screen 3 on the answer sheet.

13. Switch the Input Mode Select switch on the oscilloscope to the DC position. Draw what you see on Screen 3. Record the positive and negative peak voltage values next to the screen in the space provided.

14. Use your multimeter to measure the DC charge voltage directly across the capacitor. Record your reading on the answer sheet.

15. Turn off the generator. Answer all questions pertaining to Figure 3.

16. Turn the generator back on. With the oscilloscope still connected at Testpoints A and G, reduce the AC signal voltage and observe what happens. Check for the 0 V line by momentarily switching to the GND position of the Input Mode Select switch. Describe what you observed on the answer sheet.

17. Now, increase the AC signal voltage to a level greater than 10 V_{P-P} and observe what happens. Describe what you observe on the answer sheet.

18. Turn off all power and disassemble your circuit.

NAME:______________________________CLASS:__________DATE:__________

PART ONE QUESTIONS

Screen 1

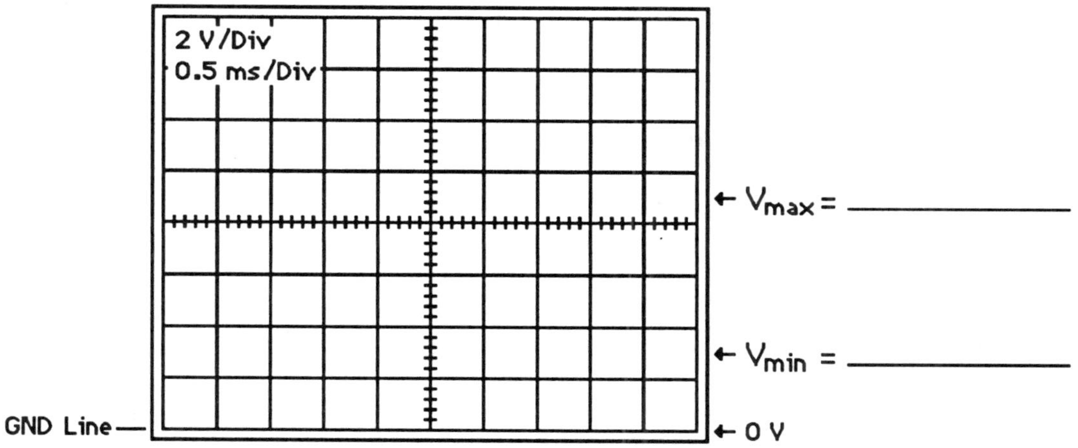

1. Is the signal displayed on Screen 1 clamped in the (a) positive direction or (b) negative direction? _______

2. Would the circuit of Figure 1 be called a (a) positive clamper or (b) negative clamper? ______

3. What would be necessary to change the circuit of Figure 1 to the opposite polarity clamper?
____
____

4. What is the reference voltage for the AC signal you drew on Screen 1? In other words, what DC voltage is the AC signal clamped to? ______________

5. What change, or changes, could you make to the circuit of Figure 1 to cause the AC signal to be clamped to +3 VDC?

PART TWO QUESTIONS

Screen 2

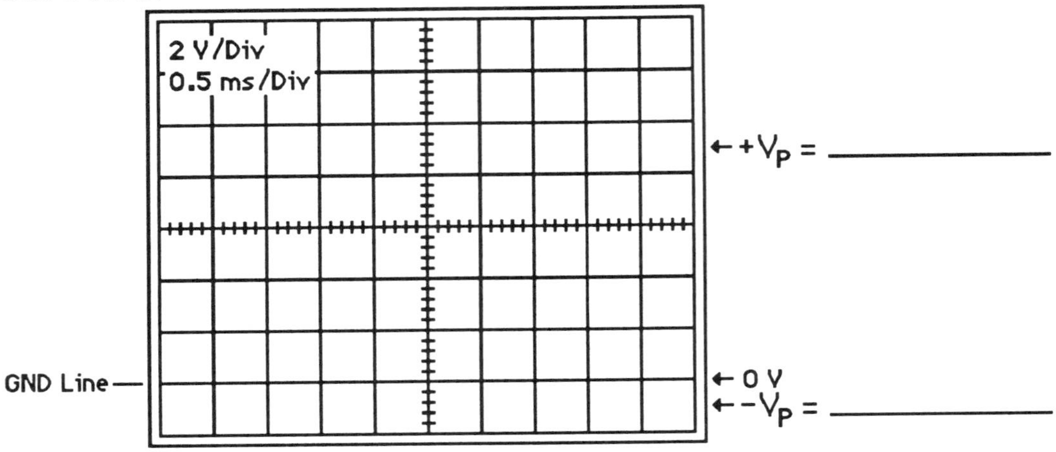

1. Is the signal displayed on Screen 2 clamped in the (a) positive direction or (b) negative direction? ________
2. Would the circuit of Figure 2 be called a (a) positive clamper or (b) negative clamper? ______
3. On what portion of the AC signal does the diode conduct? ______________________________
4. To what value of DC voltage is the AC signal of Screen 2 clamped? _______________
5. Multimeter reading: Capacitor charge voltage = _______________
6. How can you tell from the schematic whether the diode clamper is positive or negative?

____

Screen 3

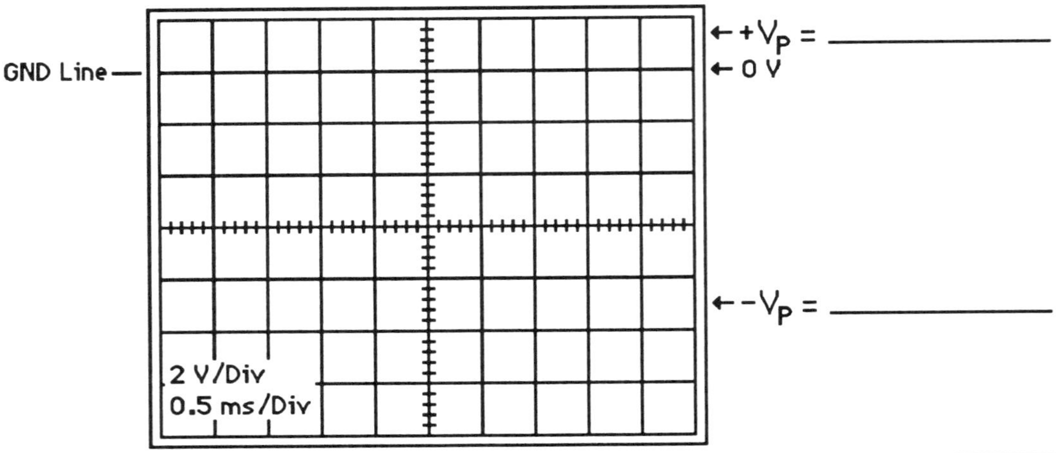

7. Is the signal displayed on Screen 3 clamped in the (a) positive direction or (b) negative direction? _______

8. Would the circuit of Figure 3 be called a (a) positive clamper or (b) negative clamper? ______

9. On what portion of the AC signal does the diode conduct? ______________________________

10. To what value of DC voltage is the AC signal of Screen 3 clamped? _______________

11. Multimeter reading: Capacitor charge voltage = ________________

12. How is the capacitor charge voltage related to the DC clamp voltage?

____

13. Describe what you observed when you decreased the signal voltage in Step 16.

____

____

14. Describe what you observed when you increased the signal voltage in Step 17.

____

____

15. Explain why not all of the AC signal is in the negative range at the output of the circuit of Figure 3. In other words, why is there a small positive peak voltage?

____

____

NOTES AND CALCULATIONS

NOTES AND CALCULATIONS

Light-Emitting Diodes

EXPERIMENT **9**

PURPOSE

The purpose of this experiment is to familiarize you with the electrical and operating characteristics of the light-emitting diode (LED).

FORMAL OBJECTIVES

After completing this experiment, you will be able to:

- describe the forward voltage and forward current characteristics of a common LED.
- utilize LEDs in circuits as status indicators.

PREPARATIONS

In preparation for this experiment, you should study the information covering light-emitting diodes in Section 3-5 of *Exploring Electronic Devices.*

COMPONENTS and EQUIPMENT

1 - Variable DC supply, 0 to 25 V
1 - Protoboard
1 - 1 kΩ, 1 W Resistor
1 - Multimeter
1 - LED (any single LED)

DISCUSSION

In the following steps, you will assemble and test a very simple light-emitting diode circuit. You will use this circuit to measure forward voltage across the LED and use the series-resistor voltage to calculate forward current where $I_f = V_{RS}/I_{RS}$. From this experiment you will see the relationship between forward LED voltage and forward current. You will be able to select a nominal forward LED voltage from a graph that you will draw of I_f vs V_f.

Figure 1

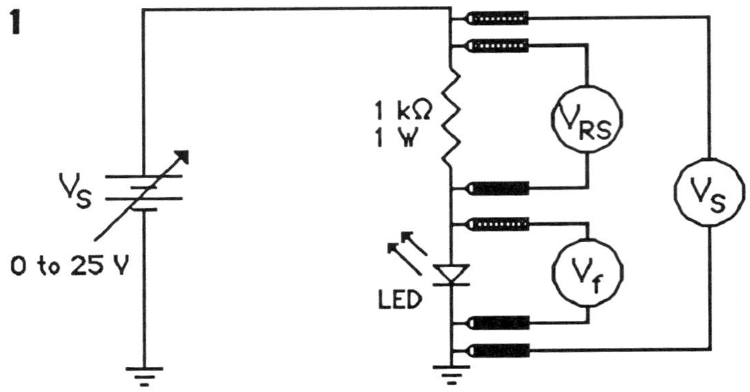

PROCEDURE

1. Identifying the terminals: Before you begin constructing your circuit, take the LED in your hand and observe the terminals and the base of the LED. Usually, one terminal lead is shorter than the other. The short lead is the negative lead (cathode). Also, next to the short lead at the base of the LED there is usually a flat area. This flat portion of the circular base also designates the cathode lead. Naturally, the longer lead is the positive lead (anode).

2. Use your ohmmeter to measure the 1-kΩ resistor. Record the exact value here. __________Ω You will use this value in your calculations for forward current where $I_f = V_{RS}/R_S$.

3. Construct the circuit shown in Figure 1.

4. Preset the DC supply to 0 V.

5. See Table 1 on the Answer sheet. Set V_S to each of the voltages specified in Table 1 and measure and record V_f and V_{RS} for each. Take care to notice at what value of V_f the LED first emits light.

   Caution: The series resistor will become very warm as voltage is increased on the high end.

6. Once you have measured and recorded all voltages in Table 1, go back and calculate the forward current for each setting of V_S.

7. Now, plot all values of V_f and I_f on Chart 1. Draw a connecting line between all plotted points to form a graph.

8. Turn off the DC supply and multimeter and answer any remaining questions.

NAME:______________________________ CLASS:__________ DATE:__________

QUESTIONS

1. At what value of V_f did the LED first emit light? ____________

2. At what values of V_f and I_f was the LED of "adequate" brightness?

   ____

3. From your graph on Chart 1, what would you say would be a nominal operating forward voltage for the LED? ____________

4. What is V_f when I_f is close to 10 mA?

   ____

5. How would you describe the graph you drew? Is it (a) linear, or (b) nonlinear?

   ______

6. How much power is the LED dissipating when the forward current is 10 mA?

   ____

7. How much power is the series resistor dissipating when the forward current is 10 mA?

   ____

8. If you decided on a nominal forward current of 10 mA and the LED was to be energized from a 12-V source, what value and wattage of series resistor would you need?
   R_S = ____________
   $P_{RS(rating)}$ = ____________

9. What value of V_f did you use in your calculations for Question 8? ____________
   Where did you get this value of V_f?

   ____
   ____
   ____
   ____

Table 1

V_S	V_f	V_{RS}	I_f
0.2 V			
0.5 V			
0.7 V			
1.0 V			
1.2 V			
1.5 V			
1.7 V			
2.0 V			
2.2 V			
2.5 V			
3.0 V			
4.0 V			
6.0 V			
9.0 V			
15 V			
20 V			
25 V			

Chart 1

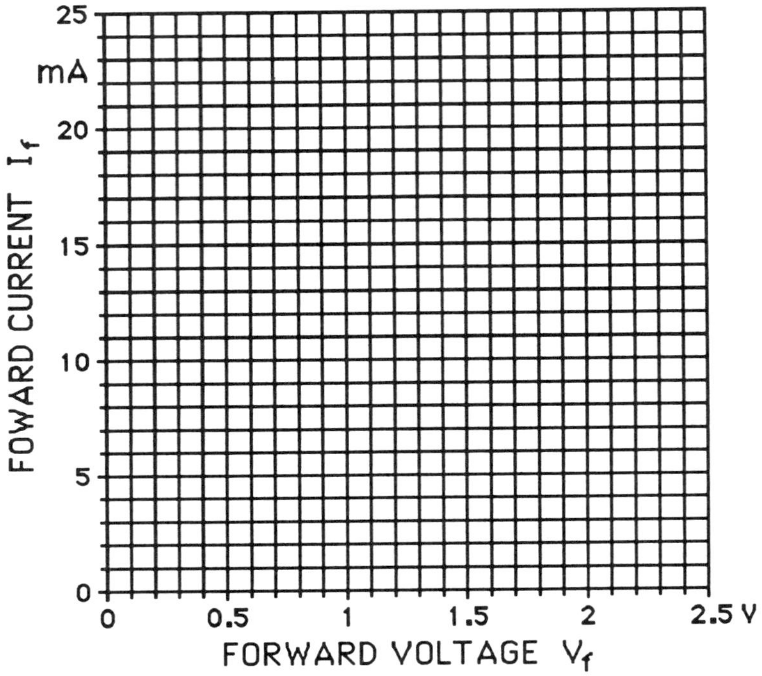

NOTES AND CALCULATIONS

Transistor Testing and Characteristics

EXPERIMENT 10

PURPOSE

The purpose of this experiment is to demonstrate difference ways of testing a transistor and to familiarize you with transistor operating characteristics.

FORMAL OBJECTIVES

After completing this experiment, you will be able to:

- use the Diode Test function of a digital multimeter to test a transistor.
- test transistors using a simple Go/No-Go test circuit.
- set up a test circuit to determine the $ß_{DC}$ of a transistor.
- set up a test circuit to develop a family of characteristic curves for a transistor.

PREPARATIONS

In preparation for this experiment, you should study Sections 4-3 and 4-4 of *Exploring Electronic Devices.*

COMPONENTS and EQUIPMENT

1 - 12 V DC Source
1 - Digital multimeter with Diode Test function
1 - Protoboard
1 - 2N3906 PNP Transistor
1 - 5 kΩ Trimmer potentiometer
2 - 47 kΩ, 1/4 W Resistor
1 - Variable DC source, 0 to 25 V
1 - 2N3904 NPN Transistor
1 - Light-emitting diode
1 - 100 kΩ, 1/4 W Resistor
1 - 1 kΩ, 1 W Resistor

PART ONE: METHODS OF TRANSISTOR TESTING

In this part of the experiment, you will explore three basic methods of testing transistors. The first method involves the use of the Diode Test function found on many digital multimeters. The second method utilizes a simple test circuit that demonstrates whether the base of the DUT (device under test) has control over the collector. It is referred to as a Go/No-Go Test. The last method we will consider is a test circuit that can be used to determine the actual DC beta of the transistor.

Using the Diode Test Function on Your Multimeter

1. Hold the 2N3904 NPN transistor in your hand so the leads are toward you and the flat side of the transistor is face up as shown in Figure 1. The leads can now be identified from left to right as emitter (E), base (B), and collector (C).

2. Set your multimeter to the Diode Test function. The red testprobe is positive and the black probe is negative.

3. Reverse Bias E to B Test: Touch the testprobes to the emitter and base lead as shown in Figure 1. The meter should read out of range indicating the junction is acting like an open. This is to be expected because the junction is now reverse biased.

Figure 1

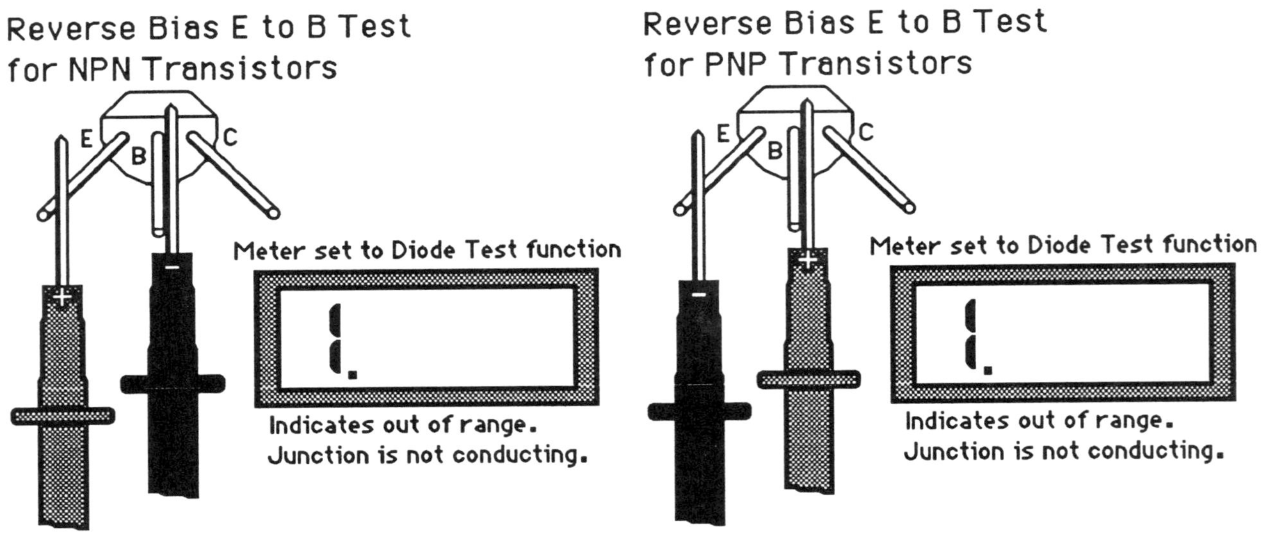

4. Forward Bias B to E Test: Next, switch the testprobes on the base and emitter leads as shown in Figure 2. The meter should read the actual forward voltage of the base-emitter junction. This indicates the junction is good. An abnormally low reading would indicate the junction is shorted and would show up in both the forward and reverse test.

5. Forward Gain Test: The last test is a simple forward gain test as shown in Figure 3. Simply slide the right probe so it touches both the base and collector at the same time. This test does not give you an actual gain reading, such as ß. It merely indicates that the transistor is conducting from collector to emitter in a biased-on condition. This is what you would expect to see when testing a good transistor. If the reading does not drop as the probe touches both the base and collector, the collector is probably open.

Figure 2

Forward Bias B to E Test
for NPN Transistors

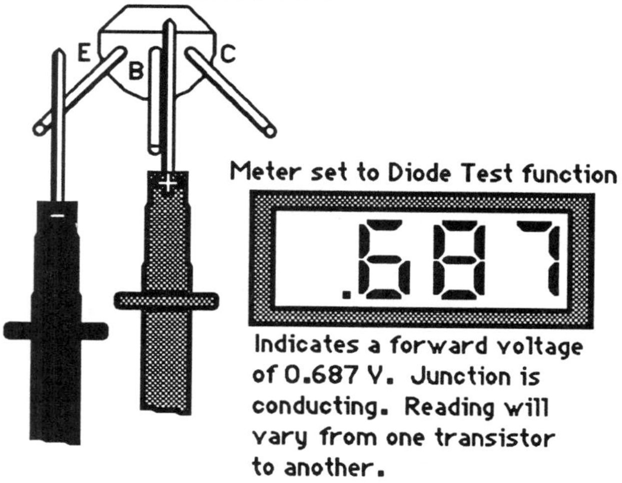

Forward Bias B to E Test
for PNP Transistors

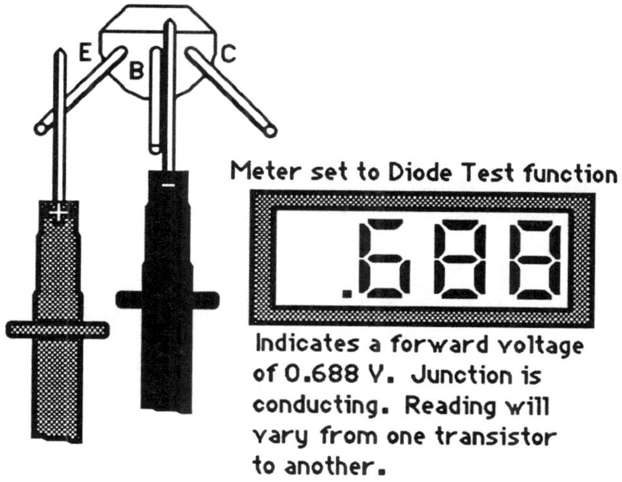

Figure 3

Forward Gain Test
for NPN Transistors

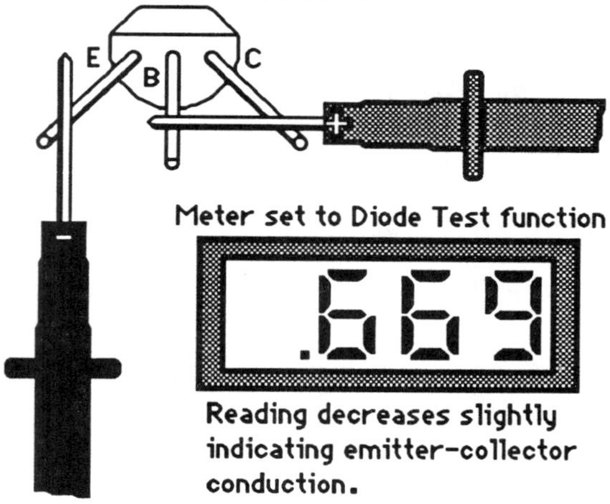

Forward Gain Test
for PNP Transistors

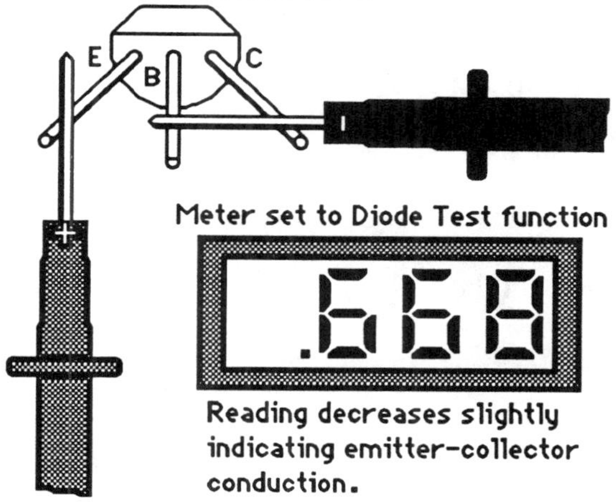

6. One additional test you can make is from the collector to the emitter with the base open. The collector to emitter should appear open regardless of which probe is on the emitter or collector as long as the base lead is not used.

7. Repeat Steps 1 through 6 with the 2N3906 PNP transistor.

Figure 4

Go/No-Go Test Circuit

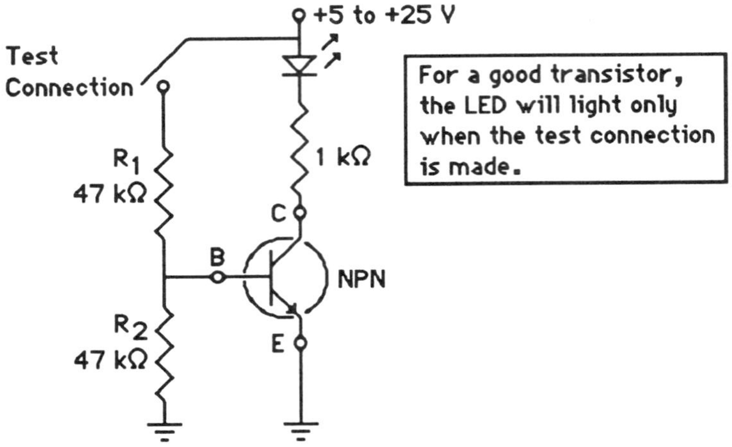

Using a Go/No-Go Test Circuit

8. Assemble the test circuit shown in Figure 4. Use a jumper wire to make the momentary test connection from the V_{CC} source to the top of R_1. Leave the jumper disconnected on one end until you are ready to test.

9. Set V_{CC} source to any voltage in the range of 5 to 20 V.

10. Momentarily touch the loose end of the jumper to the top of R_1. If the circuit is wired correctly and the transistor is good, the LED will light. This demonstrates the the base current has control over the collector current.

11. Turn off the DC power supply and disconnect the DC supply wires from your circuit.

12. Replace the 2N3904 with a 2N3906 transistor. Important: Reverse the LED in the circuit.

13. Reconnect the DC supply wires with the positive lead to circuit ground.

14. Turn on the DC supply and test the 2N3906 as you did the 2N3904. Turn off the DC supply.

Figure 5

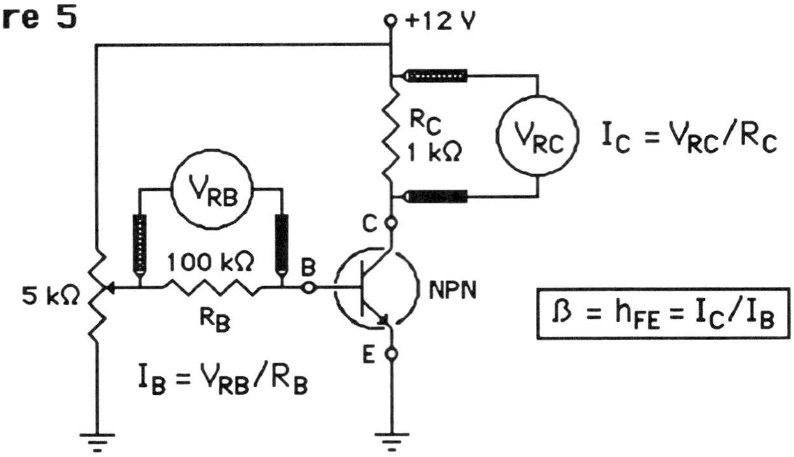

Using a Beta Test Circuit

15. Assemble the circuit shown in Figure 5 using the 2N3904 NPN transistor.

    Recall that $\beta_{DC} = h_{FE} = I_C/I_B$. Figure 5 will allow you to determine the base current and resulting collector current. This circuit will not only tell you if the transistor is good but it will allow you to determine its actual DC current gain at various values of base current.

16. Apply +12 V to the circuit as indicated.

17. Connect your voltmeter across the collector resistor R_C and adjust the trimmer potentiometer until $V_{RC} = 6$ V. $I_C = V_{RC}/R_C = 6\ \text{V}/1\ \text{k}\Omega = 6$ mA.

18. Measure the voltage across R_B and record it here. V_{RB} = 2.73v

19. Calculate the base current: $I_B = V_{RB}/R_B$ = 27.3μA

20. Now, calculate the DC current gain: $\beta_{DC} = h_{FE} = I_C/I_B$ = 6mA/27.3μA = 219.78

21. Determine the current gain for different settings of the trimmer potentiometer.

22. *Do not disassemble your circuit.* Remove power and answer all questions for Part One on the answer sheet.

PART TWO: DEVELOPING A FAMILY OF CHARACTERISTIC CURVES

In this part of the experiment, you will develop a family of characteristic collector curves for your 2N3904 transistor. These curves will graphically show the amount of saturation current for different values of base current. Later, you will use families of curves such as these to predict transistor-amplifier circuit performance. For now, it is important that you understand what these curves represent and how they are developed.

Figure 6

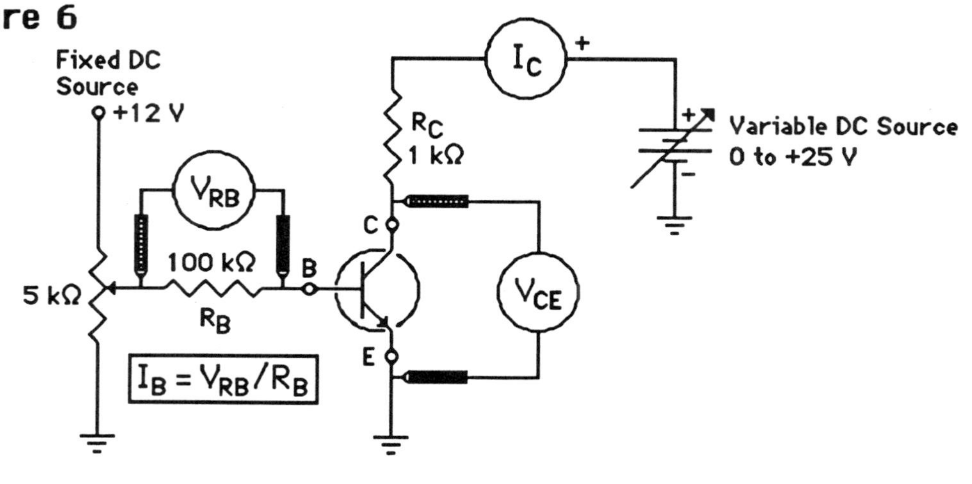

1. Modify your circuit of Figure 5 to match that of Figure 6. Notice the 5-kΩ potentiometer has a fixed +12 V DC supply. The collector supply is variable from 0 to 25 V. Also, an ammeter should be placed in series with the 1-kΩ collector resistor and the variable DC supply. If a second multimeter is not available for this purpose, you can measure voltage across the collector resistor as you did before and calculate the collector current.

2. Preset the variable DC supply to 0 V.

3. Apply power. Turn the potentiometer all the way down so $V_{RB} = 0$ V.

4. Look at Table 1 on the answer sheet. For each column, you will use the potentiometer to set V_{RB} to the designated voltage, first 2 V, then 4 V, then 6 V, etc. For each setting of V_{RB}, you will increase the transistor's collector-emitter voltage in the increments shown by carefully increasing the voltage of the variable DC supply. Record the collector current for each value of V_{CE}. Repeat this procedure for each setting of V_{RB} until the table is full.

   Caution: The transistor and the collector resistor may become very warm for higher settings of V_{RB}. Work quickly.

5. Remove power from your circuit. Turn off the DC power supply.

6. Plot the data in each column of Table 1 on Chart 1. Connect the plotted points to form a characteristic curve for each column of data. Each curve represents the relationship between collector current and collector-emitter voltage for a specific value of base current (I_C vs V_{CE} at a specific value of I_B).

7. Label each curve with its base current. For example, the bottom curve should be labeled $I_B =$ 20 μA.

8. Answer all questions for Part Two.

NAME:______________________ CLASS:__________ DATE:__________

PART ONE QUESTIONS

1. What is a DUT? Device under test

Using the Diode Test Function on Your Multimeter

2. When using the Diode Test function on your multimeter, what reading would you expect to see when testing an NPN transistor with the positive probe to the base and the negative probe to the emitter?
Should read the actual forward voltage of the base emitter junction

3. When using the Diode Test function on your multimeter, what reading would you expect to see when testing a PNP transistor with the positive probe to the collector and the negative probe to the emitter?
the meter should read out of range indicating the junction is acting like a open

4. When using the Diode Test function on your multimeter, what reading would you expect to see when testing an NPN transistor with the positive probe touching the base and collector and the negative probe touching the emitter?
indicates that the transistor is conducting from the collector to emitter in a biased on condition.

5. When using the Diode Test function on your multimeter, what reading would you expect to see when testing a PNP transistor with the positive probe to the base and the negative probe to the emitter?
should read the actual forward voltage of the base emitter junction.

Using a Go/No-Go Test Circuit

6. Briefly, explain how the Go/No-Go test circuit of Figure 4 works.
If the transistor is good the LED will light. Doesn't determine anything but whether the transistor works

7. Explain how the test circuit of Figure 4 must be set up for PNP transistors.
The same way as for NPN except that the LED must be reversed
Reverse Power -

8. What do you suppose might be wrong with the transistor if the LED lights before the test connection is made?
The diode is bad, a short between the collector and emitter

9. What do you suppose might be wrong with the transistor if the LED does not light when the test connection is made?
The transistor might be burned out, "burnt open"

Using a Beta Test Circuit

10. What change(s) must you make to the test circuit of Figure 5 in order to test PNP transistors?
Just turn the other transistor around

11. Does increasing V_{RB} cause V_{RC} to (a) increase, or (b) decrease? ____________

12. Was the β_{DC} for the 2N3904 transistor exactly the same for all settings of the trimmer potentiometer? ~~yes~~ no

PART TWO QUESTIONS

Table 1

V_{CE}	$V_{RB} = 2$ V, $I_B = 20\ \mu A$, I_C	$V_{RB} = 4$ V, $I_B = 40\ \mu A$, I_C	$V_{RB} = 6$ V, $I_B = 60\ \mu A$, I_C	$V_{RB} = 8$ V, $I_B = 80\ \mu A$, I_C	$V_{RB} = 10$ V, $I_B = 100\ \mu A$, I_C
0 V	0 A	0 A	0 A	0 A	0 A
0.1 V	.25mA	.8 mA	1.52mA	.9	2.3
0.2 V	2.1mA	4.53	7.52	.18	24.6
0.3 V	2.83mA	6.15	10.23	.23	19.7
0.4 V	2.89mA	6.52	11.34	.34	20.1
0.5 V	2.90mA	6.64	11.68	.48	21.9
1.0 V	2.93 mA	6.68	12.32	.93	23.6
1.5 V	2.96 mA	6.72	12.5	1.4	24.2
2.0 V	2.98mA	6.76	12.76	1.93	24.8
3.0 V	3.02mA	6.80	12.99	2.91	25.1
4.0 V	3.05mA	6.83	13.28	4.0	25.2
5.0 V	3.09 mA	6.87	13.58	4.92	25.5
10 V	3.22mA	6.99	14.70	9.82	25.9
15 V	3.3mA	7.09	16.26	14.56	26.2
20 V	3.52mA	7.23	17.93	17.3	26.8
25 V	3.69mA	7.29	18.76	19.5	27.1

1. How could you use the characteristic curves to determine β_{DC} for a given base current? Give an example.

____

____

____

2. What is the approximate collector-emitter voltage when the transistor first reaches its maximum value of collector current (saturation current)? $V_{CE(sat)} \sim$ .3V

Chart 1

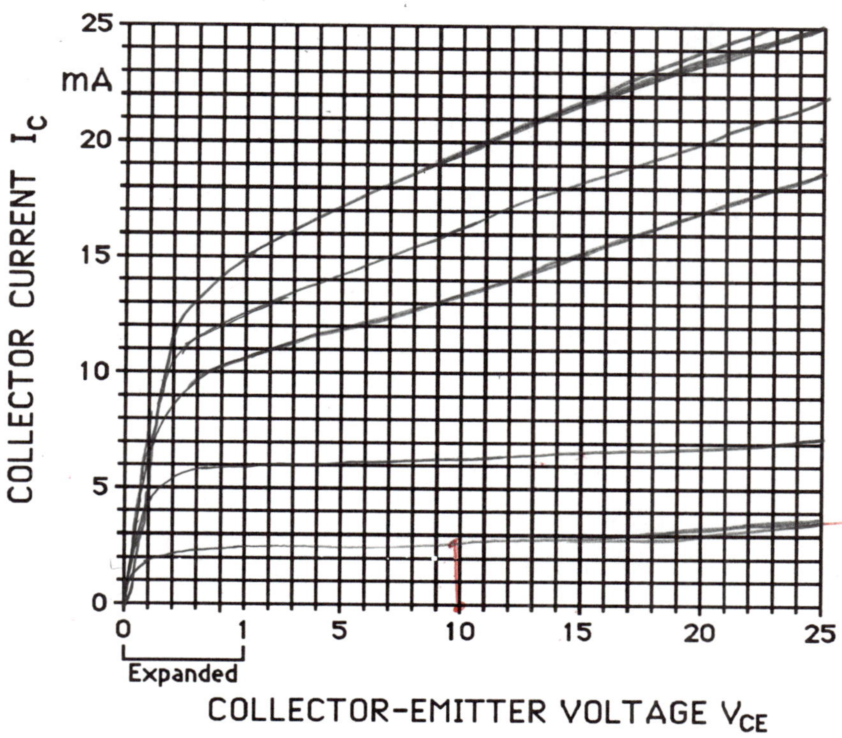

$\frac{6.5}{40} = .162$

$I_b = 20\,\mu A$

$\frac{2.8\,mA}{20\,\mu A} = \frac{2800}{20} = 140$

3. From the curves, would you say the collector current changes (a) greatly, or (b) very little as V_{CE} is varied from 1 to 25 V? b) very little

4. From the curves, what is the approximate value of collector current when I_B is 40 μA and V_{CE} is greater than 0.5 V? $I_{C(IB=40\mu A)}$ = 664

5. From the curves, would you say that collector current is related more directly to (a) base current or to (b) collector-emitter voltage? Base Explain.
A little Base voltage determines the amount of a larger current through the Collector/Emitter.

NOTES AND CALCULATIONS

Transistor Biasing Techniques

EXPERIMENT **11**

PURPOSE

The purpose of this experiment is for you to construct a variety of amplifier stages using different biasing techniques. You will do some design work and analysis. Also, you will compare the various biasing techniques in terms of temperature stability and circuit reproducibility using different transistors.

FORMAL OBJECTIVES

After completing this experiment, you will be able to:

• design, construct, and analyze base biased, collector-feedback biased, and voltage-divider biased amplifier circuits.
• construct and analyze emitter-biased amplifier circuits.
• describe the various biasing techniques in terms of temperature stability and transistor interchangeability.

PREPARATIONS

In preparation for this experiment, you should study Sections 5-1 through 5-5 of *Exploring Electronic Devices.*

COMPONENTS and EQUIPMENT

1 - Dual-polarity DC supply (±12 V)
1 - Protoboard
1 - 47 kΩ, 1/4 W Resistor
1 - 20 kΩ, 1/4 W Resistor
1 - 680 Ω, 1/4 W Resistor
1 - 150 Ω, 1/4 W Resistor
1 - 100 Ω, 1/4 W Resistor
1 - Digital multimeter
2 - 2N3904 NPN transistors
1 - 43 kΩ, 1/4 W Resistor
1 - 3.3 kΩ, 1/4 W Resistor
1 - 270 Ω, 1/4 W Resistor
1 - 120 Ω, 1/4 W Resistor
1 - 33 Ω, 1/4 W Resistor

PART ONE: BASE BIASING

In this part of the experiment, you will first construct a base-biased amplifier circuit to determine the DC beta for each of your transistors. Then, you will use the value of DC beta for one of your transistors to determine the base resistor needed to establish midpoint bias where $V_{CE(Q)}$ is approximately equal to $V_{CC}/2$. You will observe any change in the quiescent bias point caused by transistor warming. Also, the second transistor will be substituted into your base-biased circuit to see if there are any changes in the quiescent bias point.

Figure 1

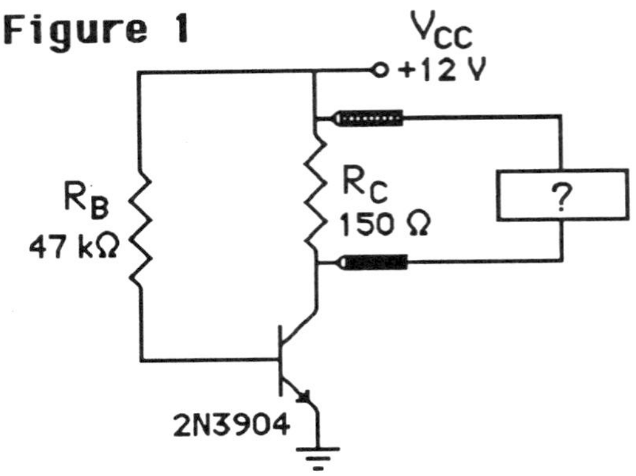

1. Assemble the circuit shown in Figure 1 and apply +12 V as indicated.
2. Measure V_{RC} and record your reading on the answer sheet in Part One.
3. In like manner, measure V_{RB} and record your reading on the answer sheet in Part One.
4. Use V_{RC} and R_C to calculate I_C. Use V_{RB} and R_B to calculate I_B. Record your calculations on the answer sheet in Part One.
5. Calculate the DC beta for your transistor and record it on the answer sheet in Part One.
6. Repeat Steps 1 through 5 with the second 2N3904 transistor. Keep these transistors separate and mark them in some way to identify their betas.

Figure 2

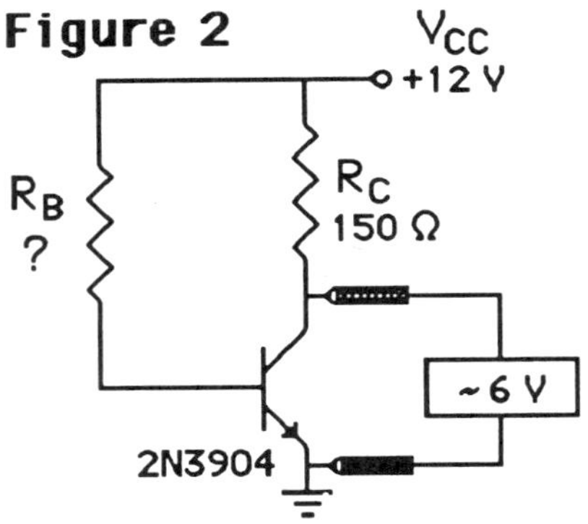

7. Select Transistor 1. Use its value of DC beta to calculate the value of R_B needed to establish midpoint bias as indicated in Figure 2 where $V_{CE(Q)} \sim 6$ V. Record your answer on the answer sheet.
8. Select a standard-value resistor close the value you calculated in Step 7 and place it in the circuit of Figure 2.
9. Measure $V_{CE(Q)}$ to see how close it is to 6 V. Record your answer on the answer sheet.
10. Remove power from your circuit and let the transistor cool.
11. Connect your voltmeter across the transistor from collector to emitter. Apply power and carefully observe the total change in $V_{CE(Q)}$ over a three minute period. As the transistor warms, the value of β_{DC} will increase causing I_C to increase and $V_{CE(Q)}$ to decrease. Record the total change in $V_{CE(Q)}$ on the answer sheet. (The total change will be in millivolts.)
12. Remove power and replace your transistor with the second 2N3904. Reapply power.

13. Measure $V_{CE(Q)}$. Record your answer on the answer sheet.

14. Remove power from your circuit. Remove R_B. Replace your transistor with the original transistor.

PART TWO: COLLECTOR-FEEDBACK BIASING

In Part Two, you will experiment with collector-feedback bias, observing temperature effects and changes in bias point caused by transistor substitution.

Figure 3

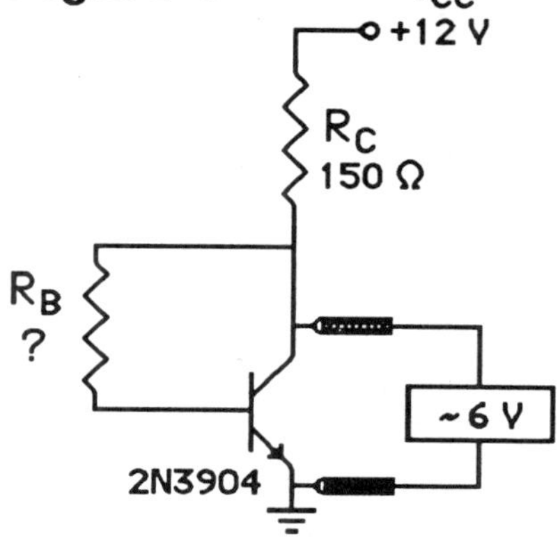

1. Calculate the value of R_B needed to establish midpoint bias for the circuit of Figure 3. Use the value of DC beta that you determined for Transistor 1. Record your answer on the answer sheet for Part Two.

2. Select a standard-value resistor close the value you calculated in Step 1 and place it in the circuit of Figure 3. Record this value on the answer sheet.

3. Measure $V_{CE(Q)}$ to see how close it is to 6 V. Record your answer on the answer sheet.

4. Remove power from your circuit and let the transistor cool.

5. Connect your voltmeter across the transistor from collector to emitter. Apply power and carefully observe the total change in $V_{CE(Q)}$ over a three minute period. As the transistor warms, the value of β_{DC} will increase causing I_C to increase and $V_{CE(Q)}$ to decrease. Record the total change in $V_{CE(Q)}$ on the answer sheet. (The total change will be in millivolts.)

6. Remove power and replace your transistor with the second 2N3904. Reapply power.

7. Measure $V_{CE(Q)}$. Record your answer on the answer sheet.

8. Remove power from your circuit. Remove R_B and R_C. Replace your transistor with the first transistor.

PART THREE: VOLTAGE-DIVIDER BIASING WITH AN EMITTER RESISTOR

In this part of the experiment, you will investigate voltage-divider bias, observing temperature effects and changes in bias point caused by transistor substitution.

1. Calculate the values of R_1 and R_2 needed to establish midpoint bias for the circuit of Figure 4. Use the value of DC beta that you determined for Transistor 1. Record your answer on the answer sheet for Part Three. (Review the design procedure given in Section 5-3 of *Exploring Electronic Devices*.)

2. Select standard-value resistors close the values you calculated in Step 1 and place them in the circuit of Figure 4. Record these values on the answer sheet.

3. Measure $V_{CE(Q)}$ to see how close it is to 6 V. Record your answer on the answer sheet.

4. Remove power from your circuit and let the transistor cool.

5. Connect your voltmeter across the transistor from collector to emitter. Apply power and carefully observe the total change in $V_{CE(Q)}$ over a three minute period. As the transistor warms, the value of β_{DC} will increase causing I_C to increase and $V_{CE(Q)}$ to decrease. Record the total change in $V_{CE(Q)}$ on the answer sheet. (The total change will be in millivolts.)

6. Remove power and replace your transistor with the second 2N3904. Reapply power.

7. Measure $V_{CE(Q)}$. Record your answer on the answer sheet.

Figure 4

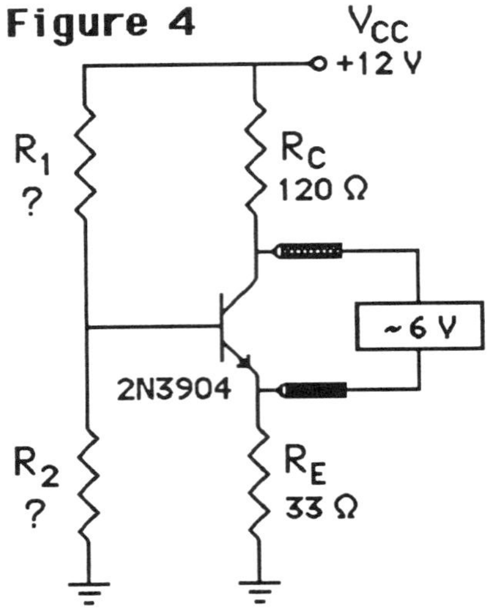

8. Remove power from your circuit. Remove R_C, R_E, R_1, and R_2. Replace your transistor with the first transistor.

PART FOUR: EMITTER BIASING

In this part of the experiment, you will explore emitter biasing, observing temperature effects and changes in bias point caused by transistor substitution.

Figure 5

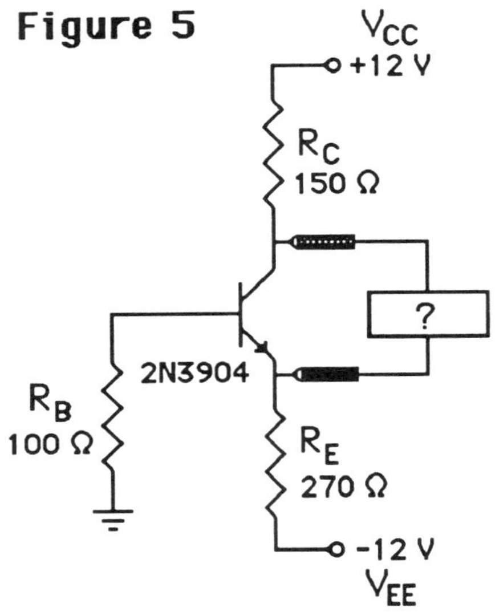

1. Assemble the circuit shown in Figure 5. Use Transistor 1. Do not apply power at this time.

2. Calculate the predicted values of $I_{C(Q)}$ and $V_{CE(Q)}$ and record them on the answer sheet for Part Four.

3. Use your multimeter to measure $I_{C(Q)}$ and $V_{CE(Q)}$ and record them on the answer sheet for Part Four.

4. Measure V_{RB}, V_E, and V_C. Record your readings on the answer sheet for Part Four.

5. Remove power from your circuit and let the transistor cool.

6. Connect your voltmeter across the transistor from collector to emitter. Apply power and carefully observe the total change in $V_{CE(Q)}$ over a three minute period. As the transistor warms, the value of β_{DC} will increase causing I_C to increase and $V_{CE(Q)}$ to decrease. Record the total change in $V_{CE(Q)}$ on the answer sheet. (The total change will be in millivolts.)

7. Remove power and replace your transistor with the second 2N3904. Reapply power.
8. Measure $V_{CE(Q)}$. Record your answer on the answer sheet.
9. Remove power and disassemble your circuit.

NOTES AND CALCULATIONS

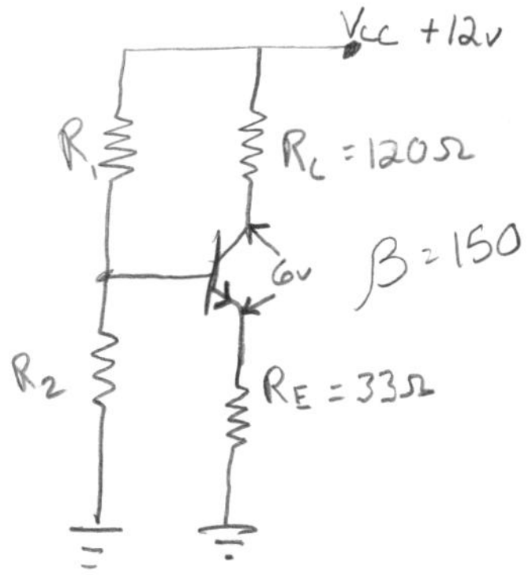

$V_E = .7v$

$V_B = 1.4v$

$R_2 \leq V_{R_2}/I_{R_2}$

$R_1 = (V_{CC} - V_{R_2})/I_{R_2} + I_B$

$V_{R_2} = 1.99v$

$I_C = 39.2mA$

$I_B = 261\mu A$

$I_{R_2} = 2.6mA$

NAME:__________ **CLASS:**__________ **DATE:**__________

PART ONE QUESTIONS

1. For Figure 1, Transistor 1: V_{RC} = 7.37v, V_{RB} = 11.29v
   Transistor 2: V_{RC} = 5.39v, V_{RB} = 11.26v

2. For Figure 1, Transistor 1: I_C = 49.1mA, I_B = 240μA
   Transistor 2: I_C = 35.9mA, I_B = 240μA

3. Transistor 1: β_{DC} = 204.58
   Transistor 2: β_{DC} = 149.58

4. Use Transistor 1: For Figure 2, R_B should be 57.79kΩ Ω.
   Show calculations here.

   40mA

5. Value of R_B actually used in Figure 2 is 56kΩ Ω.

6. Using Transistor 1: $V_{CE(Q)}$ = 5.37v

7. Total change in $V_{CE(Q)}$ as temperature increased was 370mV mV.

8. Using Transistor 2: $V_{CE(Q)}$ = 7.7v

9. Describe how increasing temperature effects the transistor's beta, collector current, and collector-emitter voltage.

   β went down, V_{CE} went down, collector current went down

10. What happened to $V_{CE(Q)}$ when the second transistor was placed in the circuit of Figure 2? Why?

    the voltage was higher because the β was not the same as the other transistor.

PART TWO QUESTIONS

1. For Figure 3, R_B should be 19.8kΩ Ω.
   Show your calculations here.

2. Value of R_B actually used in Figure 3 is 20 k Ω.

3. Using Transistor 1: $V_{CE(Q)}$ = 5.1v

4. Total change in $V_{CE(Q)}$ as temperature increased was 70 mV.

5. Using Transistor 2: $V_{CE(Q)}$ = 6.24v

6. What happened to $V_{CE(Q)}$ when the second transistor was placed in the circuit of Figure 3? Why?

The voltage is higher on the second resistor Because the B is lower than the first.

PART THREE QUESTIONS

1. For Figure 4, R_1 = 760 Ω, and R_2 = 3.5k Ω.
Show calculations here.

2. Actual values chosen for R_1 and R_2: R_1 = 750 Ω, and R_2 = 3.6k Ω

3. Using Transistor 1: $V_{CE(Q)}$ = 5.95v

4. Total change in $V_{CE(Q)}$ as temperature increased was 10 mV.

5. Using Transistor 2: $V_{CE(Q)}$ = ~~6.00~~ 5.99v

6. What happened to $V_{CE(Q)}$ when the second transistor was placed in the circuit of Figure 4? Why?

The voltage increased slightly because the Beta is higher.

PART FOUR QUESTIONS

1. For Figure 5, the predicted values are: $I_{C(Q)}$ = 42mA, $V_{CE(Q)}$ = 6.4v
Show your calculations here.

$V_{CE} = V_{CC} + V_E - V_{RC} - V_{RE}$ $\quad I_{CQ} = \frac{11.3v}{270\Omega}$

$12 + 12 \quad 6.3 - 11.3v$

$24 - 6.3 - 11.3$

$= 6.4v$

2. Measured values: $I_{C(Q)}$ = 37.5mA, $V_{CE(Q)}$ = 6.3v

3. Measured values: V_{RB} = 6.9v, V_E = -.7v, V_C = 7.0v

4. Do the calculated values in Question 1 exactly match the measured values recorded in Question 2? Why?
Close

5. Does $V_{CE(Q)}$ recorded in Question 2 equal $V_C - V_E$? Show mathematically here.
7.0v - .7v = 6.3v yes

6. Total change in $V_{CE(Q)}$ as temperature increased was 87 mV.

7. List the different biasing techniques in order of stability from the most temperature stable to the least stable.

1. Emitter Bias (most stable)
2. Voltage Divider
3. Collector FeedBack
4. Base Bias (least stable)

8. Which of the biasing techniques was the one that showed the least change in quiescent bias point when the second transistor was placed in the circuit? Why?
Emitter Bias - Not Dependent on Beta

The Common-Emitter Amplifier

EXPERIMENT 12

PURPOSE

The purpose of this experiment is to give you the opportunity to compare the results of mathematical analysis of a common-emitter amplifier with the results of experimental analysis. In so doing, you will be reassured of the validity of the theoretical analysis presented in *Exploring Electronic Devices.*

FORMAL OBJECTIVES

After completing this experiment, you will be able to:

- use a DMM (DVM) to measure quiescent DC voltages of a common-emitter amplifier.
- use an oscilloscope to measure input and output signal voltage to determine the amplification factor.
- use a variable resistor along with an oscilloscope to determine the input and output impedance of a common-emitter amplifier.

PREPARATIONS

In preparation for this experiment, you should study Section 6-1 of *Exploring Electronic Devices* and the appendix of this manual titled *Using the Oscilloscope.*

COMPONENTS and EQUIPMENT

1 - DC source (12 V, 10 mA capability)
1 - Plug-in experiment board (breadboard)
1 - Dual-trace oscilloscope with test probes
1 - 100 μF, 25 V Electrolytic capacitor
1 - 50 kΩ Trimmer potentiometer
1 - 68 kΩ, 1/4 W Resistor
1 - 4.7 kΩ, 1/4 W Resistor
1 - 220 Ω, 1/4 W Resistor
1 - Digital multimeter with clip leads
1 - Function generator with clip leads
1 - 2N3904 NPN Transistor (or equivalent)
2 - 10 μF, 25 V Electrolytic capacitors
1 - 10 kΩ Trimmer potentiometer
1 - 10 kΩ, 1/4 W Resistor
1 - 470 Ω, 1/4 W Resistor

PART ONE: THEORETICAL ANALYSIS

In this part of the experiment, you will perform a mathematical (theoretical) analysis of a common-emitter BJT amplifier.

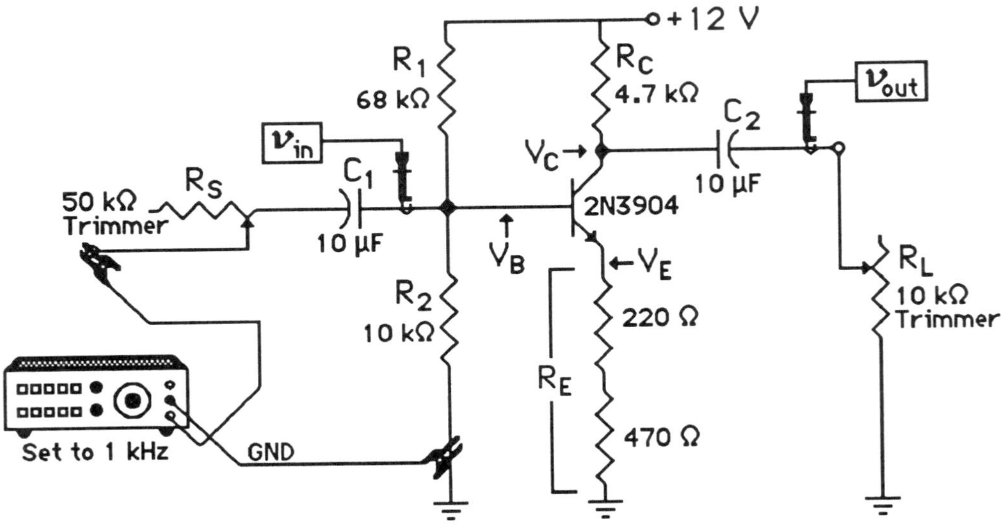

Figure 1

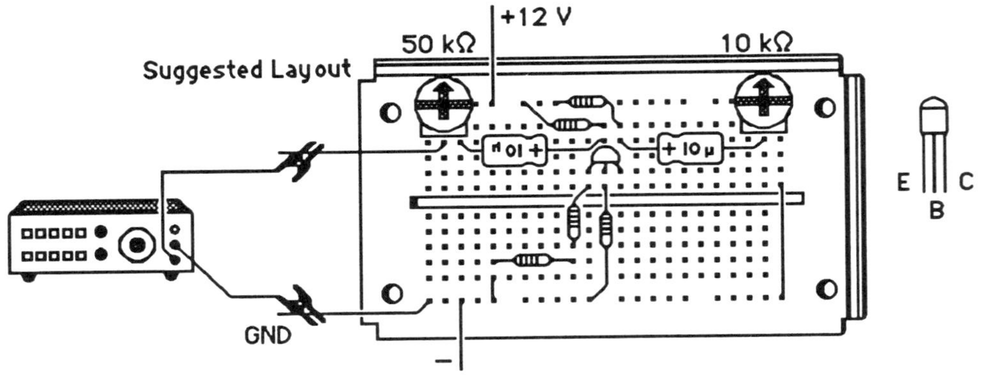

1. Perform a complete mathematical analysis of the amplifier shown in Figure 1. Record your results on the answer sheet by filling in all blanks for Part One including the first column of Table 1.

2. For this experiment, you will use an NPN transistor whose β_{DC} you know from Experiment 10. Record its β_{DC} where requested on the answer sheet for Part One.

3. Determine and draw the DC and AC load lines for the circuit of Figure 1. Mark the Q point. The AC load line should be drawn assuming a 10-kΩ load resistance.

PART TWO: EXPERIMENTAL ANALYSIS USING TEST EQUIPMENT

In this part of the experiment, you will use the DMM (DVM) and the oscilloscope along with a function generator to determine DC and AC parameters. Carefully construct the common-emitter amplifier circuit shown in Figure 1. *For now, leave the 10-kΩ trim pot out of the circuit.* Ask someone to verify the accuracy of your construction. Apply 12 V to your circuit as indicated in Figure 1 when you are sure it is wired correctly.

Record all measurements or calculations in Table 1 where indicated. Compare measured values with calculated values. They should be close.

1. Measuring V_B: With no test signal applied from the function generator, use a DVM to measure the voltage from the base to ground.

2. Measuring V_C: With no test signal applied from the function generator, use a DVM to measure the voltage from the collector to ground.

3. Measuring V_E: With no test signal applied from the function generator, use a DVM to measure the voltage from the emitter to ground.

4. Measuring V_{CE}: With no test signal applied from the function generator, use a DVM to measure the voltage from the collector to the emitter of the transistor. This is the quiescent collector to emitter voltage.

5. Setting v_{out} to 1 $V_{P\text{-}P}$: Set the function generator to produce a 1 kHz sine wave. Preset the function generator's amplitude control to minimum (fully counterclockwise). Use a clip-lead cable to connect the signal to the wiper-arm terminal of the 50-kΩ input trim pot. Set the trim pot so its resistance is 0 Ω. Connect Channel 2 of the oscilloscope to the output of the amplifier (v_{out}). Set the input mode select for Channel 2 to the AC position. Set the volts/div control to 0.5 V/div. Set the trigger for Auto and trigger from Channel 2. Set the time/div control for 0.5 ms/div. Turn on the function generator and the oscilloscope. Slowly increase the amplitude control on the function generator until the output voltage shown on the oscilloscope is 1 $V_{P\text{-}P}$. Use the Channel 2 vertical position knob to place this signal in the lower half of the screen.

6. Measuring v_{in}: Connect Channel 1 of the oscilloscope to the input of the amplifier at the transistor's base (v_{in}). Set the input mode select for Channel 1 to the AC position. Set the volts/div control for Channel 1 as appropriate to measure the peak-to-peak input voltage. Use the Channel 1 vertical position knob to place this signal in the upper half of the screen.

7. Determining $A_{V(no\ load)}$: Using the measured input and output voltages, calculate the amplifier's amplification factor ($A_{V(no\ load)}$).

8. Input-output phase relationship: Observe and record the phase relationship of the input and output AC signals displayed on the screen.

9. Determining $v_{out\ (max,\ no\ load)}$: Increase the function generator's amplitude control (clockwise) while observing the amplifier's output voltage on Channel 2. Adjust the Channel 2 volts/div control as needed. Stop when the peak-to-peak output reaches maximum without any clipping of the positive or negative peak.

10. Measuring $v_{in\ (10\ k\Omega\ load)}$: Place the 10 kΩ trimmer potentiometer in the circuit as shown in Figure 1 and set it for maximum resistance (10 kΩ). Adjust the function generator's amplitude control as needed to achieve a 1 $V_{P\text{-}P}$ output signal (Channel 2, lower trace). Set the volts/div control for Channel 1 as appropriate to measure the peak-to-peak input voltage.

11. Determining $A_{V(RL = 10\ k\Omega)}$: Using the measured input voltage and 1-$V_{P\text{-}P}$ output voltage, calculate the amplifier's amplification factor.

12. Determining $v_{out\ (max,\ RL = 10\ k\Omega)}$: Increase the function generator's amplitude control (clockwise) while observing the amplifier's output voltage on Channel 2. Adjust the Channel 2 volts/div control as needed. Stop when the peak-to-peak output reaches maximum without any clipping of the positive or negative peak.

13. Determining Z_{in}: Adjust the function generator's amplitude control so the amplifier's output voltage is once again 1 $V_{P\text{-}P}$. Increase the resistance of the 50-kΩ trimmer potentiometer from 0 Ω until the input and output voltages drop in half ($v_{out} = 0.5\ V_{P\text{-}P}$). Now disconnect the function generator from the wiper arm of the 50-kΩ trim pot and use an ohmmeter to measure the resistance of the trim pot. The resistance of the trim pot from its wiper arm terminal to its C_1 terminal will equal the input impedance of the amplifier. Reconnect the function generator to the wiper terminal of the 50-kΩ trim pot.

14. Determining Z_{out}: Make sure the resistance of the 50-kΩ trim pot is reset to 0 Ω. Temporarily remove the 10-kΩ trim pot and adjust the function generator's amplitude control to once again obtain an output voltage of 1 $V_{P\text{-}P}$. Now reinsert the 10-kΩ trim pot and adjust its resistance so the output voltage drops to 0.5 $V_{P\text{-}P}$. The amplifier's output voltage is now being evenly divided between its output impedance and the resistance of the trim pot. Disconnect C_2 from the 10-kΩ trim pot and use your ohmmeter to measure the resistance of the trim pot from wiper arm to ground. This resistance is equal to the output impedance of the amplifier.

PART THREE: INCREASING GAIN WITH AN EMITTER-BYPASS CAPACITOR

In this part of the experiment, you will investigate the effects of adding a 100-µF emitter-bypass capacitor to your circuit as shown in Figure 2. You will make DC and AC measurements as you did in Part Two and record your findings in Column 3 of Table 1 for comparison. Follow the guidelines presented in Part Two to determine each parameter.

1. Add the 100-µF emitter-bypass capacitor to your amplifier circuit as shown in Figure 2.

2. Follow the guidelines presented in Part Two to determine each parameter and record them in Column 3 of Table 1.

3. Answer all questions on the answer sheet.

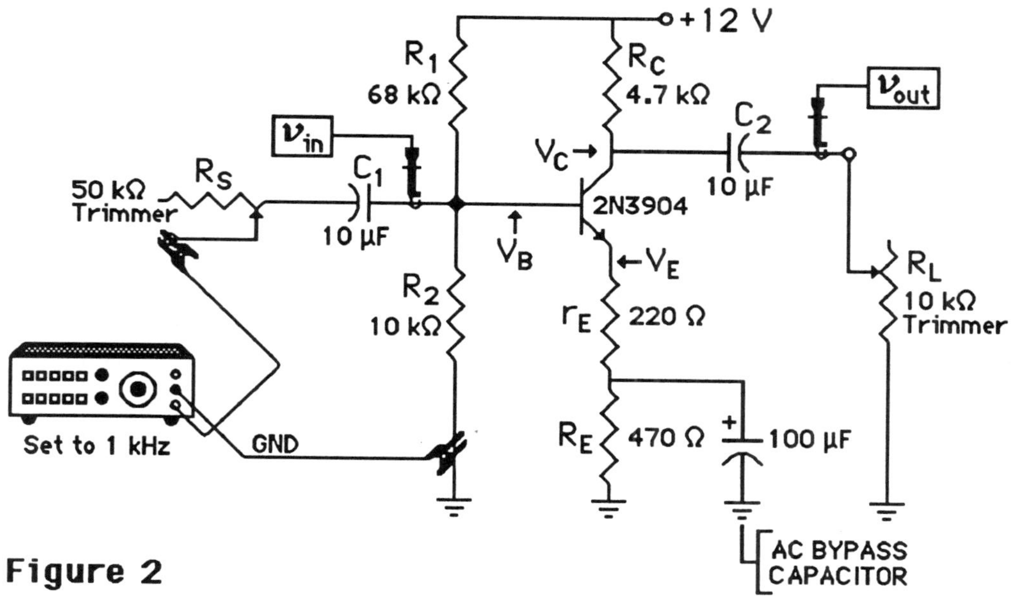
+12 V
R_1
68 kΩ
R_C
4.7 kΩ
v_{out}
C_2
v_{in}
V_C
R_S
50 kΩ
Trimmer
C_1
10 μF
2N3904
10 μF
V_B
V_E
R_L
R_2
10 kΩ
r_E
220 Ω
10 kΩ
Trimmer
Set to 1 kHz
GND
R_E
470 Ω
100 μF
AC BYPASS
CAPACITOR

Figure 2

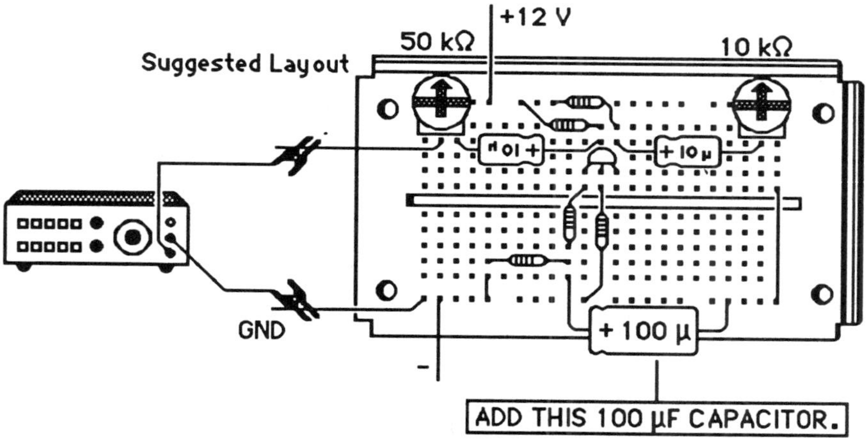
+12 V
50 kΩ
10 kΩ
Suggested Layout
.01 +
+ 10 μ
GND
+ 100 μ
ADD THIS 100 μF CAPACITOR.

NAME:____________________________________CLASS:__________DATE:__________

PART ONE QUESTIONS

1. β_{DC} = ______________

2. Draw and label the DC and AC load lines for your amplifier on Chart 1. The AC load line should takes into consideration the 10 kΩ load resistor but no emitter-bypass capacitor.

Chart 1

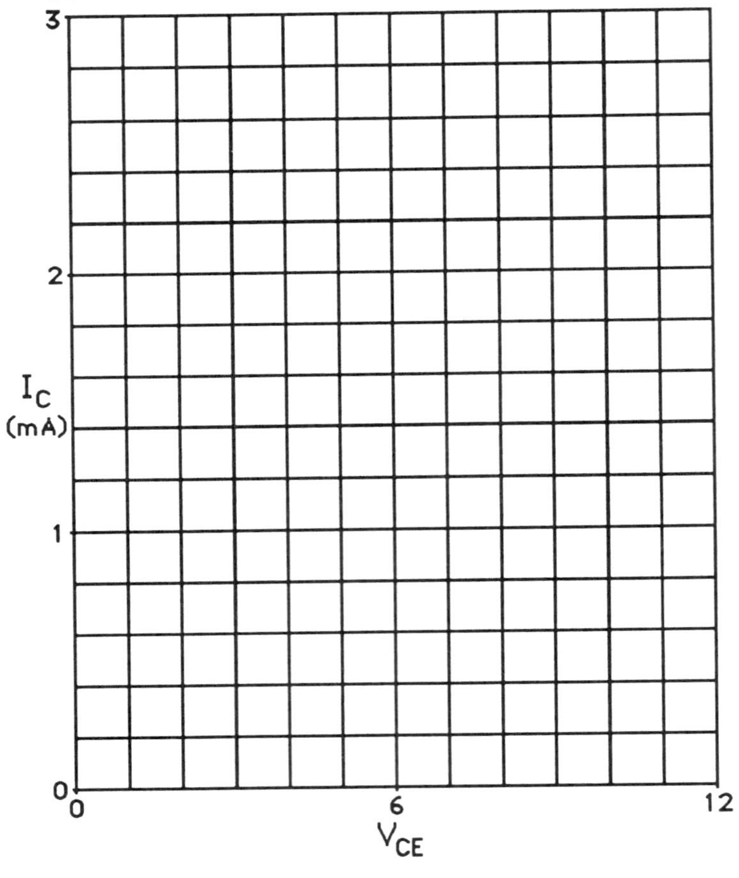

TABLE 1

	PARAMETER	CALCULATED (Part One)	MEASURED (Part Two)	MEASURED (Part Three)
1.	V_B =			
2.	V_C =			
3.	V_E =			
4.	V_{CE} =			
5.	v_{out} =	1 $V_{P\text{-}P}$	1 $V_{P\text{-}P}$	1 $V_{P\text{-}P}$
6.	v_{in} =			
7.	$A_{V(\text{no load})}$ =			
8.	v_{in} to v_{out} Phase Shift =			
9.	$v_{out\ (\text{max, no load})}$ =			
10.	v_{in} (RL = 10 kΩ) =			
11.	$A_{V(RL = 10\ k\Omega)}$ =			
12.	$v_{out\ (\text{max, RL = 10 k}\Omega)}$ =			
13.	Z_{in} =			
14.	Z_{out} =			

PART THREE QUESTIONS

1. From the data collected in Table 1, describe the effect the 10-kΩ load resistor has on voltage gain (A_V).

   ____

   ____

2. From the data collected in Table 1, describe the effect the 10-kΩ load resistor has on maximum output voltage.

   ____

   ____

3. From the data collected in Table 1, describe what effect the 100-μF emitter-bypass capacitor has on voltage gain (A_V).

   ____

   ____

4. From the data collected in Table 1, describe what effect the 100-µF emitter-bypass capacitor has on input impedance (Z_{in}).

5. List the parameters that are not affected by load resistance.

NOTES AND CALCULATIONS

The Common-Collector Amplifier

EXPERIMENT 13

PURPOSE

The purpose of this experiment is to give you the opportunity to compare the results of mathematical analysis of a common-collector amplifier with the results of experimental analysis. In so doing, you will be reassured of the validity of the theoretical analysis presented in *Exploring Electronic Devices.*

FORMAL OBJECTIVES

After completing this experiment, you will be able to:

- use a DMM (DVM) to measure quiescent DC voltages of a common-collector amplifier.
- use an oscilloscope to measure input and output signal voltage to determine the amplification factor.

PREPARATIONS

In preparation for this experiment, you should study Section 6-2 of *Exploring Electronic Devices* and the appendix of this manual titled *Using the Oscilloscope.*

COMPONENTS and EQUIPMENT

1 - DC source (12 V, 20 mA capability)
1 - Plug-in experiment board (breadboard)
1 - Dual-trace oscilloscope with test probes
2 - 10 μF, 35 V Electrolytic capacitors
1 - 3.3 kΩ, 1/4 W Resistor
1 - Digital multimeter with clip leads
1 - Function generator with clip leads
1 - 2N3904 NPN Transistor (or equivalent)
2 - 10 kΩ, 1/4 W Resistor
1 - 1 kΩ, 1/4 W Resistor

PART ONE: THEORETICAL ANALYSIS

In this part of the experiment, you will perform a mathematical (theoretical) analysis of a common-collector BJT amplifier.

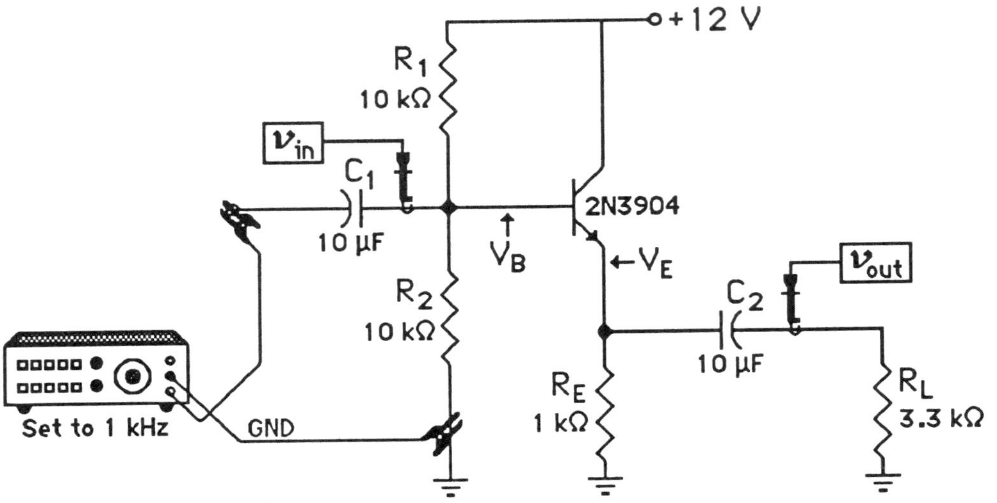

Figure 1

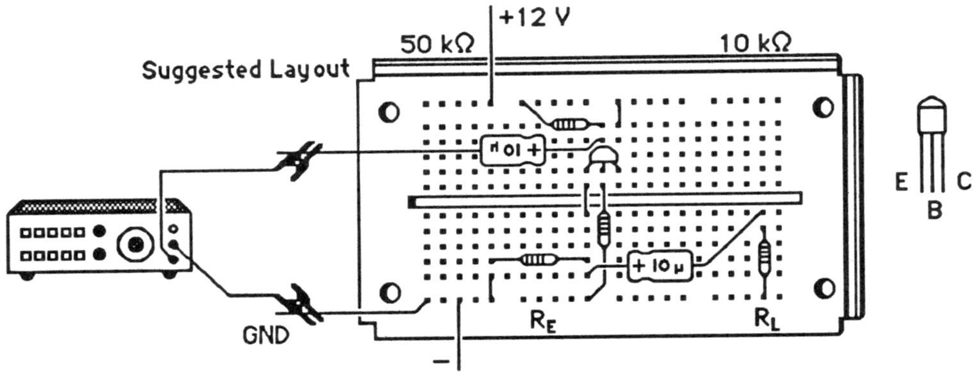

1. Perform a complete mathematical analysis of the amplifier shown in Figure 1. Record your results on the answer sheet by filling in all blanks for Part One including the first column of Table 1.

2. For this experiment, you will use an NPN transistor whose β_{DC} you know from Experiment 10. Record its β_{DC} where requested on the answer sheet for Part One.

3. Determine and draw the DC and AC load lines for the circuit of Figure 1. Mark the Q point. The AC load line will demonstrate the effect of the 3.3-kΩ load resistance.

PART TWO: EXPERIMENTAL ANALYSIS USING TEST EQUIPMENT

In this part of the experiment, you will use the DMM (DVM) and the oscilloscope along with a function generator to determine DC and AC parameters. Carefully construct the common-collector amplifier circuit shown in Figure 1. *For now, leave the 3.3-kΩ load resistor out of the circuit.* Ask someone to verify the accuracy of your construction. Apply 12 V to your circuit as indicated in Figure 1 when you are sure it is wired correctly.

Record all measurements or calculations in Table 1 where indicated. Compare measured values with calculated values. They should be very close.

1. Measuring V_B: With no test signal applied from the function generator, use a DVM to measure the voltage from the base to ground.

2. Measuring V_E: With no test signal applied from the function generator, use a DVM to measure the voltage from the emitter to ground.

3. Measuring V_{CE}: With no test signal applied from the function generator, use a DVM to measure the voltage from the collector to the emitter of the transistor. This is the quiescent collector to emitter voltage.

4. Setting v_{out} to 1 $V_{P\text{-}P}$ (no load): Set the function generator to produce a 1 kHz sine wave. Preset the function generator's amplitude control to minimum (fully counterclockwise). Use a clip-lead cable to connect the signal to the input coupling capacitor. Connect Channel 2 of the oscilloscope to the output of the amplifier (v_{out}). Set the input mode select for Channel 2 to the AC position. Set the volts/div control to 0.5 V/div. Set the trigger for Auto and trigger from Channel 2. Set the time/div control for 0.5 ms/div. Turn on the function generator and the oscilloscope. Slowly increase the amplitude control on the function generator until the output voltage shown on the oscilloscope is 1 $V_{P\text{-}P}$. Use the Channel 2 vertical position knob to place this signal in the lower half of the screen.

5. Measuring v_{in}: Connect Channel 1 of the oscilloscope to the input of the amplifier at the transistor's base (v_{in}). Set the input mode select for Channel 1 to the AC position. Set the volts/div control for Channel 1 as appropriate to measure the peak-to-peak input voltage. Use the Channel 1 vertical position knob to place this signal in the upper half of the screen.

6. Determining $A_{V(\text{no load})}$: Using the measured input and output voltages, calculate the amplifier's amplification factor ($A_{V(\text{no load})}$).

7. Input-output phase relationship: Observe and record the phase relationship of the input and output AC signals displayed on the screen.

8. Determining $v_{out\,(\text{max, no load})}$: Increase the function generator's amplitude control (clockwise) while observing the amplifier's output voltage on Channel 2. Adjust the Channel 2 volts/div control as needed. Stop when the peak-to-peak output reaches maximum without any clipping of the positive or negative peak.

9. Measuring $v_{in\ (3.3\ k\Omega\ load)}$: Place the 3.3 kΩ load resistor in the circuit as shown in Figure 1. Adjust the function generator's amplitude control as needed to achieve a 1 $V_{P\text{-}P}$ output signal (Channel 2, lower trace). Set the volts/div control for Channel 1 as appropriate to measure the peak-to-peak input voltage.

10. Determining $A_{V(RL = 3.3\ k\Omega)}$: Using the measured input voltage and 1-$V_{P\text{-}P}$ output voltage, calculate the amplifier's amplification factor.

11. Determining $v_{out\ (max,\ RL = 3.3\ k\Omega)}$: Increase the function generator's amplitude control (clockwise) while observing the amplifier's output voltage on Channel 2. Adjust the Channel 2 volts/div control as needed. Stop when the peak-to-peak output reaches maximum without any clipping of the positive or negative peak.

12. Answer all questions on the answer sheet.

NAME:______________________________ CLASS:__________ DATE:__________

PART ONE QUESTIONS

1. β_{DC} = ______________

2. Draw and label the DC and AC load lines for your amplifier on Chart 1. The AC load line should takes into consideration the 3.3 kΩ load resistor.

Chart 1

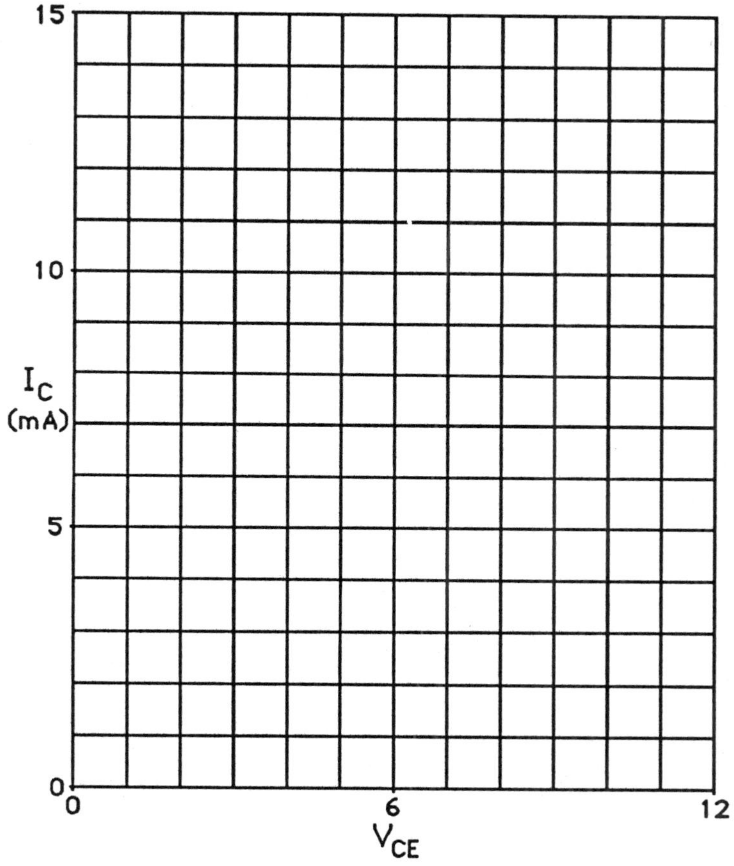

TABLE 1

	PARAMETER	CALCULATED (Part One)	MEASURED (Part Two)
1.	V_B =	____________	____________
2.	V_E =	____________	____________
3.	V_{CE} =	____________	____________
4.	v_{out} =	1 $V_{P\text{-}P}$	1 $V_{P\text{-}P}$
5.	v_{in} =	____________	____________
6.	$A_{V(\text{no load})}$ =	____________	____________
7.	v_{in} to v_{out} Phase Shift =	____________	____________
8.	$v_{out\ (\text{max, no load})}$ =	____________	____________
9.	$v_{in\ (RL = 3.3\ k\Omega)}$ =	____________	____________
10.	$A_{V(RL = 3.3\ k\Omega)}$ =	____________	____________
11.	$v_{out\ (\text{max, } RL = 3.3\ k\Omega)}$ =	____________	____________

PART TWO QUESTIONS

1. From the data collected in Table 1, describe the effect a 3.3-kΩ load resistance has on voltage gain (A_V).

____

____

2. From the data collected in Table 1, describe the effect a 3.3-kΩ load resistor has on maximum output voltage.

____

____

The Common-Base Amplifier

EXPERIMENT 14

PURPOSE

The purpose of this experiment is to give you the opportunity to compare the results of mathematical analysis of a common-base amplifier with the results of experimental analysis. In so doing, you will be reassured of the validity of the theoretical analysis presented in *Exploring Electronic Devices.*

FORMAL OBJECTIVES

After completing this experiment, you will be able to:

- use a DMM (DVM) to measure quiescent DC voltages of a common-base amplifier.
- use an oscilloscope to measure input and output signal voltage to determine the amplification factor.
- use a variable resistor along with an oscilloscope to determine the input and output impedance of a common-base amplifier.

PREPARATIONS

In preparation for this experiment, you should study Section 6-3 of *Exploring Electronic Devices* and the appendix of this manual titled *Using the Oscilloscope.*

COMPONENTS and EQUIPMENT

1 - DC source (12 V, 10 mA capability)
1 - Plug-in experiment board (breadboard)
1 - Dual-trace oscilloscope with test probes
2 - 10 µF, 35 V Electrolytic capacitors
1 - 1 kΩ Trimmer potentiometer
1 - 22 kΩ, 1/4 W Resistors
2 - 4.7 kΩ, 1/4 W Resistors
1 - Digital multimeter with clip leads
1 - Function generator with clip leads
1 - 2N3904 NPN Transistor (or equivalent)
1 - 100 µF, 50 V Electrolytic capacitor
1 - 10 kΩ Trimmer potentiometer
1 - 10 kΩ, 1/4 W Resistor

PART ONE: THEORETICAL ANALYSIS

In this part of the experiment, you will perform a mathematical (theoretical) analysis of a common-base BJT amplifier.

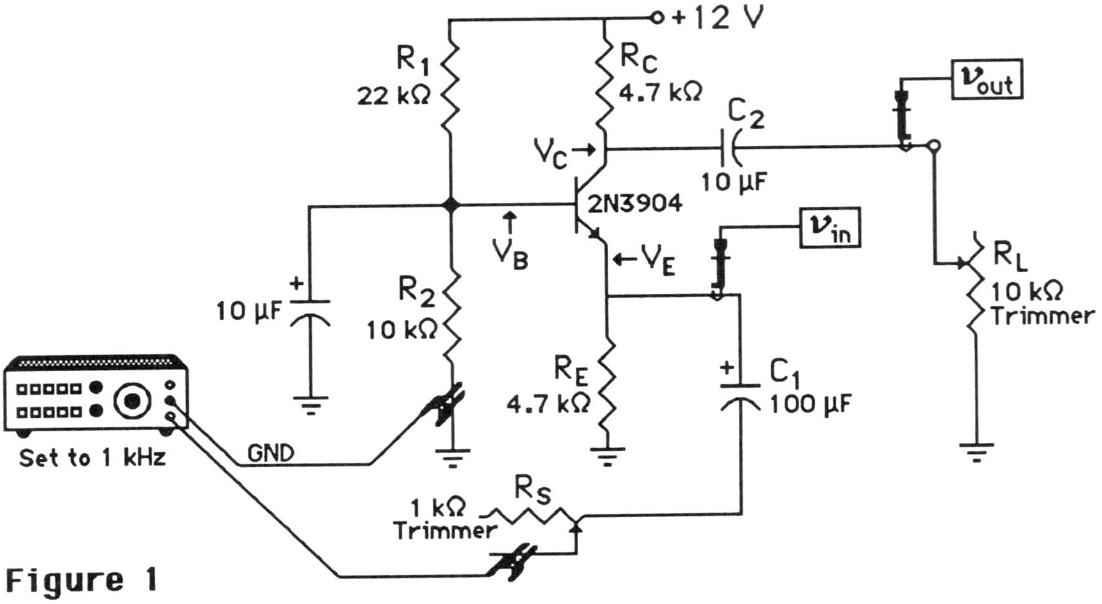

Figure 1

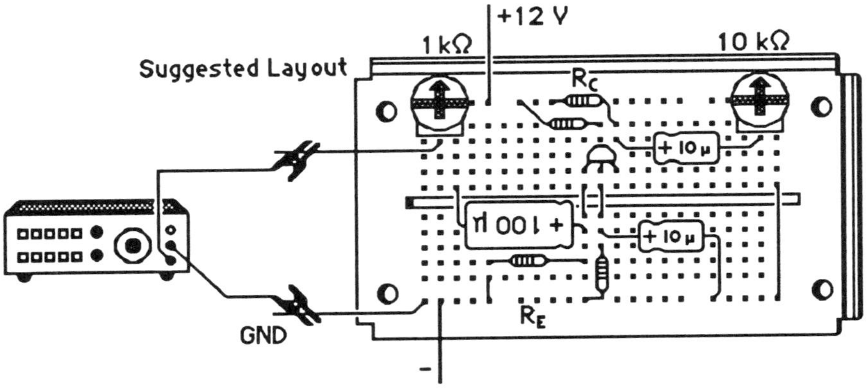

1. Perform a complete mathematical analysis of the amplifier shown in Figure 1. Record your results on the answer sheet by filling in all blanks for Part One including the first column of Table 1.

2. For this experiment, you will use an NPN transistor whose β_{DC} you know from Experiment 10. Record its β_{DC} where requested on the answer sheet for Part One.

3. Determine and draw the DC and AC load lines for the circuit of Figure 1. Mark the Q point. The AC load line should be drawn assuming a 10-kΩ load resistance.

PART TWO: EXPERIMENTAL ANALYSIS USING TEST EQUIPMENT

In this part of the experiment, you will use the DMM (DVM) and the oscilloscope along with a function generator to determine DC and AC parameters. Carefully construct the common-base amplifier circuit shown in Figure 1. *For now, leave the 10-kΩ trim pot out of the circuit.* Ask someone to verify the accuracy of your construction. Apply 12 V to your circuit as indicated in Figure 1 when you are sure it is wired correctly.

Record all measurements or calculations in Table 1 where indicated. Compare measured values with calculated values. They should be close.

1. Measuring V_B: With no test signal applied from the function generator, use a DVM to measure the voltage from the base to ground.

2. Measuring V_C: With no test signal applied from the function generator, use a DVM to measure the voltage from the collector to ground.

3. Measuring V_E: With no test signal applied from the function generator, use a DVM to measure the voltage from the emitter to ground.

4. Measuring V_{CE}: With no test signal applied from the function generator, use a DVM to measure the voltage from the collector to the emitter of the transistor. This is the quiescent collector to emitter voltage.

5. Setting v_{out} to 1 V_{P-P}: Set the function generator to produce a 1 kHz sine wave. Preset the function generator's amplitude control to minimum (fully counterclockwise). Use a clip-lead cable to connect the signal to the wiper-arm terminal of the 1-kΩ input trim pot. Set the trim pot so its resistance is 0 Ω. Connect Channel 2 of the oscilloscope to the output of the amplifier (v_{out}). Set the input mode select for Channel 2 to the AC position. Set the volts/div control to 0.5 V/div. Set the trigger for Auto and trigger from Channel 2. Set the time/div control for 0.5 ms/div. Turn on the function generator and the oscilloscope. Slowly increase the amplitude control on the function generator until the output voltage shown on the oscilloscope is 1 V_{P-P}. Use the Channel 2 vertical position knob to place this signal in the lower half of the screen.

6. Measuring v_{in}: Connect Channel 1 of the oscilloscope to the input of the amplifier at the transistor's base (v_{in}). Set the input mode select for Channel 1 to the AC position. Set the volts/div control for Channel 1 as appropriate to measure the peak-to-peak input voltage. Use the Channel 1 vertical position knob to place this signal in the upper half of the screen.

7. Determining $A_{V(no\ load)}$: Using the measured input and output voltages, calculate the amplifier's amplification factor ($A_{V(no\ load)}$).

8. Input-output phase relationship: Observe and record the phase relationship of the input and output AC signals displayed on the screen.

9. Determining $v_{out\ (max,\ no\ load)}$: Increase the function generator's amplitude control (clockwise) while observing the amplifier's output voltage on Channel 2. Adjust the Channel 2 volts/div control as needed. Stop when the peak-to-peak output reaches maximum without any clipping of the positive or negative peak.

10. Measuring $v_{in\ (10\ k\Omega\ load)}$: Place the 10 kΩ trimmer potentiometer in the circuit as shown in Figure 1 and set it for maximum resistance (10 kΩ). Adjust the function generator's amplitude control as needed to achieve a 1 $V_{P\text{-}P}$ output signal (Channel 2, lower trace). Set the volts/div control for Channel 1 as appropriate to measure the peak-to-peak input voltage.

11. Determining $A_{V(RL\,=\,10\ k\Omega)}$: Using the measured input voltage and 1-$V_{P\text{-}P}$ output voltage, calculate the amplifier's amplification factor.

12. Determining $v_{out\ (max,\ RL\,=\,10\ k\Omega)}$: Increase the function generator's amplitude control (clockwise) while observing the amplifier's output voltage on Channel 2. Adjust the Channel 2 volts/div control as needed. Stop when the peak-to-peak output reaches maximum without any clipping of the positive or negative peak.

13. Determining Z_{in}: Adjust the function generator's amplitude control so the amplifier's output voltage is once again 1 $V_{P\text{-}P}$. Increase the resistance of the 1-kΩ trimmer potentiometer from 0 Ω until the input and output voltages drop in half (v_{out} = 0.5 $V_{P\text{-}P}$). Now disconnect the function generator from the wiper arm of the 1-kΩ trim pot and use an ohmmeter to measure the resistance of the trim pot. The resistance of the trim pot from its wiper arm terminal to its C_1 terminal will equal the input impedance of the amplifier. Reconnect the function generator to the wiper terminal of the 1-kΩ trim pot.

14. Determining Z_{out}: Make sure the resistance of the 1-kΩ trim pot is reset to 0 Ω. Temporarily remove the 10-kΩ trim pot and adjust the function generator's amplitude control to once again obtain an output voltage of 1 $V_{P\text{-}P}$. Now reinsert the 10-kΩ trim pot and adjust its resistance so the output voltage drops to 0.5 $V_{P\text{-}P}$. The amplifier's output voltage is now being evenly divided between its output impedance and the resistance of the trim pot. Disconnect C_2 from the 10-kΩ trim pot and use your ohmmeter to measure the resistance of the trim pot from wiper arm to ground. This resistance is equal to the output impedance of the amplifier.

15. Answer all questions on the answer sheet.

NAME:______________________________ CLASS:__________ DATE:__________

PART ONE QUESTIONS

1. β_{DC} = ______________

2. Draw and label the DC and AC load lines for your amplifier on Chart 1. The AC load line should takes into consideration the 10 kΩ load resistor but no emitter-bypass capacitor.

Chart 1

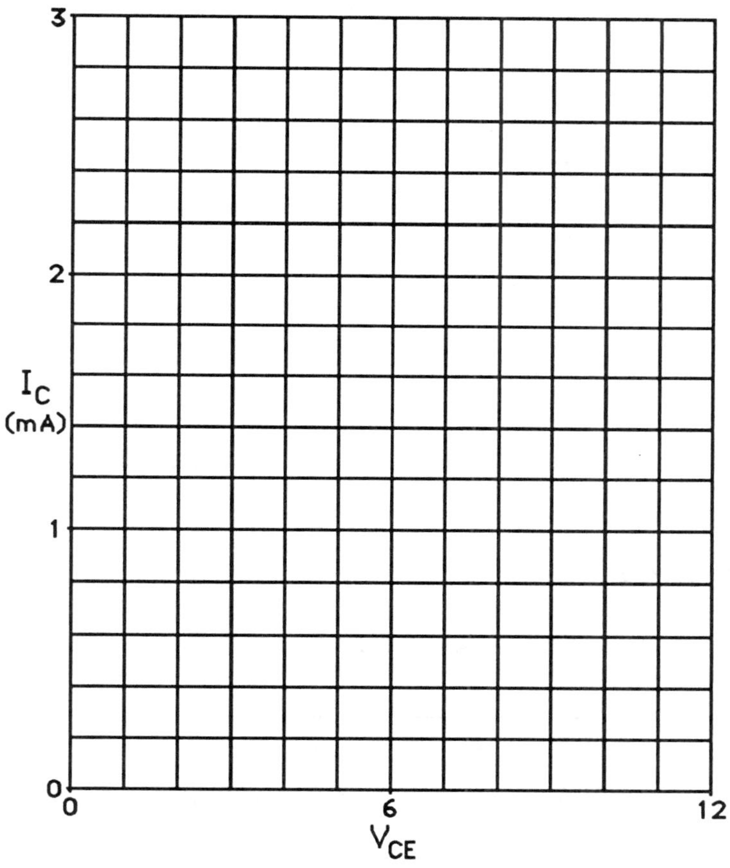

TABLE 1

	PARAMETER	CALCULATED (Part One)	MEASURED (Part Two)
1.	V_B =	________	________
2.	V_C =	________	________
3.	V_E =	________	________
4.	V_{CE} =	________	________
5.	v_{out} =	1 $V_{P\text{-}P}$	1 $V_{P\text{-}P}$
6.	v_{in} =	________	________
7.	$A_{V(no\ load)}$ =	________	________
8.	v_{in} to v_{out} Phase Shift =	________	________
9.	$v_{out\ (max,\ no\ load)}$ =	________	________
10.	$v_{in\ (RL = 10\ k\Omega)}$ =	________	________
11.	$A_{V(RL = 10\ k\Omega)}$ =	________	________
12.	$v_{out\ (max,\ RL = 10\ k\Omega)}$ =	________	________
13.	Z_{in} =	________	________
14.	Z_{out} =	________	________

PART TWO QUESTIONS

1. From the data collected in Table 1, describe the effect a 10-kΩ load resistance has on voltage gain (A_V).

____

____

2. From the data collected in Table 1, describe the effect a 10-kΩ load resistor has on maximum output voltage.

____

____

3. Explain why the input impedance is so low (demonstrate mathematically).

____

____

____

____

Class A Amplifiers with Troubleshooting

EXPERIMENT 15

PURPOSE

The purpose of this experiment is to give you the opportunity to observe the results of various troubles in class A amplifiers. You will see how a variety of troubles affects both DC and AC parameters.

FORMAL OBJECTIVES

After completing this experiment, you will be able to:

- use a DMM (DVM) and an oscilloscope to determine class A amplifier troubles.
- relate various class A amplifier troubles to possible causes.

PREPARATIONS

In preparation for this experiment, you should study Sections 6-1, 6-4, and 7-1 of *Exploring Electronic Devices* .

COMPONENTS and EQUIPMENT

1 - DC source (12 V, 10 mA capability)
1 - Plug-in experiment board (breadboard)
1 - Dual-trace oscilloscope with test probes
1 - 100 µF, 50 V Electrolytic capacitor
2 - 68 kΩ, 1/4 W Resistors
1 - 4.7 kΩ, 1/4 W Resistor
1 - 220 Ω, 1/4 W Resistor
1 - Digital multimeter with clip leads
1 - Function generator with clip leads
1 - 2N3904 NPN Transistor (or equivalent)
2 - 10 µF, 35 V Electrolytic capacitors
2 - 10 kΩ, 1/4 W Resistors
1 - 470 Ω, 1/4 W Resistor

PART ONE: THEORETICAL AND EXPERIMENTAL ANALYSIS

In this part of the experiment, you will perform a mathematical and experimental analysis of a class A amplifier.

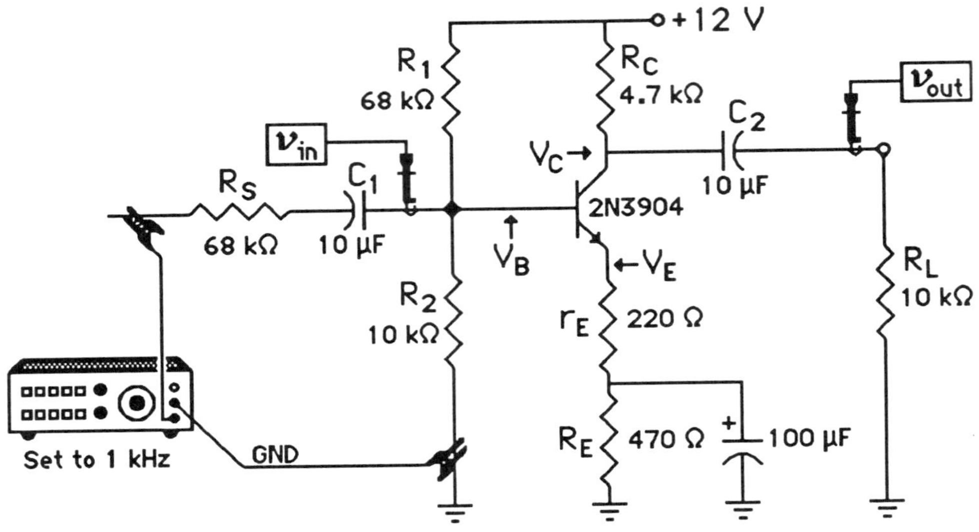

Figure 1

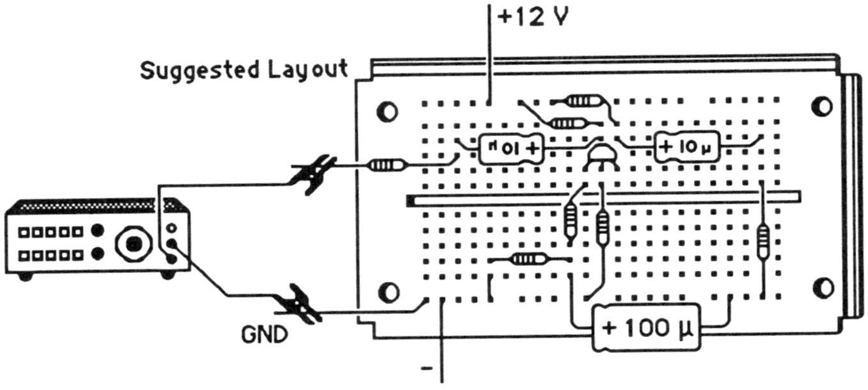

1. Perform a mathematical analysis of the amplifier shown in Figure 1. Record your results on the answer sheet by filling in all requested blanks in the first row of Table 1.

2. For this experiment, you will use an NPN transistor whose β_{DC} you know from Experiment 10. Record its β_{DC} where requested on the answer sheet for Part One.

3. Carefully construct the circuit of Figure 1. Use your test equipment, as in previous experiments, to determine actual DC and AC parameters. Do not apply the test signal when measuring DC parameters. When determining the AC input voltage and voltage gain, adjust the amplitude control on the signal generator to produce a 1 $V_{P\text{-}P}$ output signal. Record your readings in the second row of Table 1.

PART TWO: TROUBLES AND TROUBLESHOOTING

In this part of the experiment, you will use the DMM (DVM) and the oscilloscope along with a function generator to determine DC and AC parameters with a variety of troubles.

To complete Table 1, you will simulate the troubles listed in column 1 using a jumper wire to short some components. Open components can be simulated by unpluging one of its leads from the protoboard. As you complete the table, try to predict the results of each trouble before you make your measurements. Discuss the measured results of each trouble with your lab partner or fellow classmate.

For all DC parameters, the signal generator must be disconnected. For AC parameters, connect the signal generator as shown in Figure 1. The signal generator's amplitude control should still be set as it was in Part One of this experiment. For a normal untroubled amplifier, the output would be 1 V_{P-P}.

Also, as you know, A_V is not a measured parameter - it must be calculated after the input and output signal voltages are measured.

NAME:______________________ CLASS:__________ DATE:__________

QUESTIONS

1. β_{DC} = ______________

2. Row 1 of Table 1 is for calculated values and row 2 of Table 1 is for measured values with no troubles inserted.

TABLE 1

TROUBLE	V_B	V_E	V_C	V_{CE}	v_{in}	v_{out}	A_V
1. NONE	______	______	______	______	______	______	______
2. NONE	______	______	______	______	______	______	______
3. R_2 Shorted	______	______	______	______	______	______	______
4. R_2 Open	______	______	______	______	______	______	______
5. R_C Shorted	______	______	______	______	______	______	______
6. R_C Open	______	______	______	______	______	______	______
7. R_L Shorted	______	______	______	______	______	______	______
8. R_L Open	______	______	______	______	______	______	______
9. R_E Shorted	______	______	______	______	______	______	______
10. R_E Open	______	______	______	______	______	______	______
11. r_E Shorted	______	______	______	______	______	______	______
12. r_E Open	______	______	______	______	______	______	______
13. C_2 Open	______	______	______	______	______	______	______
14. C_2 Shorted	______	______	______	______	______	______	______
15. Emitter bypass capacitor is open.	______	______	______	______	______	______	______
16. Transistor shorted from collector to base.	______	______	______	______	______	______	______
17. Transistor open at the collector.	______	______	______	______	______	______	______
18. Transistor open at its base.	______	______	______	______	______	______	______
19. Transistor shorted base to emitter.	______	______	______	______	______	______	______
20. Transistor shorted collector to emitter.	______	______	______	______	______	______	______

Class B Amplifiers

EXPERIMENT 16

PURPOSE

The purpose of this experiment is to familiarize you with two common class B amplifiers, the common-emitter and the complementary symmetry push-pull amplifier. In this experiment you will prove class B theory.

FORMAL OBJECTIVES

After completing this experiment, you will be able to:

- troubleshoot basic class B amplifiers using the DMM and oscilloscope.
- describe class B characteristics and operation.

PREPARATIONS

In preparation for this experiment, you should study Section 7-2 of *Exploring Electronic Devices*.

COMPONENTS and EQUIPMENT

1 - DC source (+12 V, +18 V)
1 - Plug-in experiment board (breadboard)
1 - Function generator
1 - 2N3906 PNP transistor
2 - 100 μF, 50 V Electrolytic capacitors
1 - 18 kΩ, 1/4 W Resistor
2 - 1 kΩ, 1/4 W Resistors
1 - 680 Ω, 1/4 W Resistor
1 - Analog or digital multimeter with leads
1 - Dual-trace oscilloscope with probes
1 - 2N3904 NPN transistor
2 - 1N914 Switching diodes
2 - 10 μF, 35 V Electrolytic capacitors
2 - 10 kΩ, 1/4 W Resistors
1 - 270 Ω, 1/4 W Resistor

PART ONE: THE CLASS B COMMON-EMITTER AMPLIFIER

In this part of the experiment, you will construct and test a class B common-emitter amplifier. You should recall from your reading that class B amplifiers operate with a 50% duty cycle. That means they only amplify approximately half of the input waveform. In the case of the common-emitter configuration, the transistor is biased near cutoff so $V_{CE(Q)}$ is nearly equal to V_{CC} which places the collector close to V_{CC}. Since the output is taken from the collector, as shown in Figure 1, the collector voltage can only decrease. Thus, only the positive alternation of the input signal can be amplified due to the 180° phase inversion. The negative alternation of the input signal only forces the transistor deaper into cutoff.

Figure 1

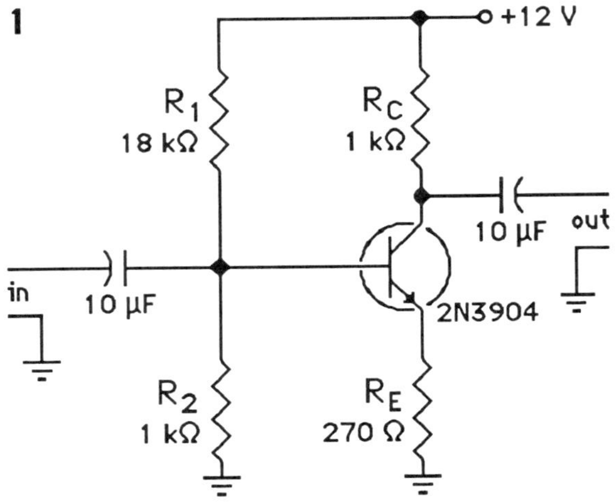

1. Assemble the circuit shown in Figure 1.
2. Apply V_{CC} (+12 V) and use your DMM to measure the DC parameters listed on the answer sheet for Part One. Record your readings.
3. Connect Channel 1 of your oscilloscope to the input of your amplifier and Channel 2 to the output. Use the upper half of the screen for Channel 1 and the lower half for Channel 2. The input mode select for each channel should be set to the AC position. Trigger from Channel 2.
4. Connect your function generator to the input and set the frequency to 1 kHz (sine wave). Increase the amplitude control of the generator slowly and observe input and output signals on the oscilloscope. You should see severe clipping on the positive alternation of the output signal. This is because the collector of the transistor starts from near VCC since the transistor is nearly cutoff. The collector voltage cannot increase any further. It can only decrease. Thus, the negative alternation of the output is unclipped.
5. Remove power from your circuit and answer all questions for Part One on the answer sheet.

PART TWO: THE COMPLEMENTARY SYMMETRY PUSH-PULL AMPLIFIER

In this part of the experiment, you will investigate the very efficient and common push-pull amplifier as shown in Figure 2. As you can see, each of the complementary transistors is configured as a class B common-collector amplifier. That means these transistors are functioning as current and power amplifiers, not voltage amplifiers. Each amplifier is biased at threshold using a diode for each base-emitter junction.

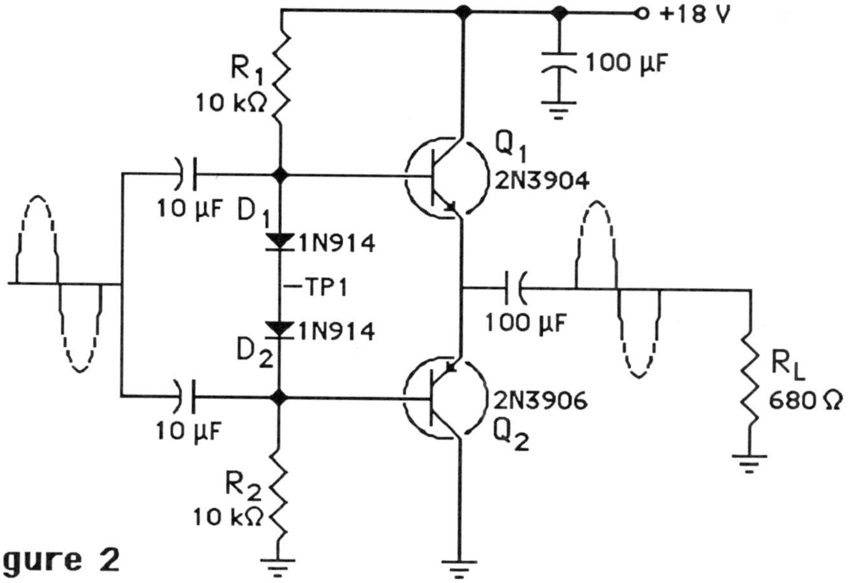

Figure 2

Suggested Layout

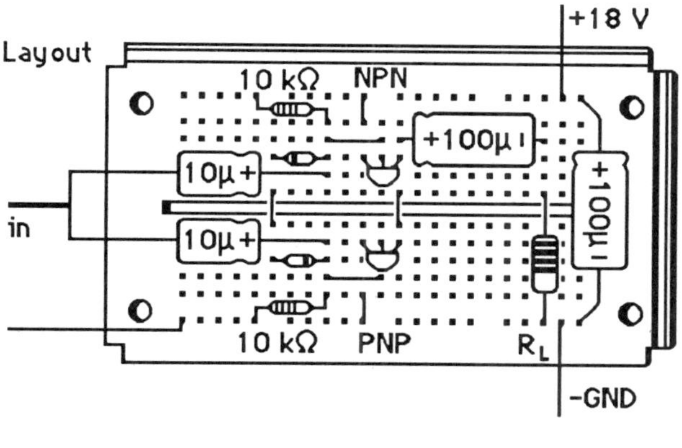

1. Assemble the circuit shown in Figure 2.

2. Apply V_{CC} (+18 V) and use your DMM to measure the DC parameters listed on the answer sheet for Part One. Record your readings.

3. Connect Channel 1 of your oscilloscope to the input of your amplifier and Channel 2 to the output. Use the upper half of the screen for Channel 1 and the lower half for Channel 2. The input mode select for each channel should be set to the AC position. Trigger from Channel 2.

4. Connect your function generator to the input and set the frequency to 1 kHz (sine wave). Increase the amplitude control of the generator slowly and observe input and output signals on the oscilloscope.

5. Set the input signal level so the output is at its maximum unclipped level. Record this maximum level on the answer sheet.

6. Measure and record the DC current that is being supplied to the amplifier under this maximum-signal condition.

7. Reduce the input signal level to 3 $V_{P\text{-}P}$. Short each diode one at a time using a small jumper wire. Observe and describe what you see on the oscilloscope.

8. Remove power from your circuit and make sure you answer all questions for Part Two on the answer sheet.

NOTES AND CALCULATIONS

NAME:______________________ CLASS:__________ DATE:__________

PART ONE QUESTIONS

1. V_B = __________, V_E = __________, V_C = __________, $V_{CE(Q)}$ = __________

   $I_{C(Q)}$ = __________

2. What is the phase relationship between the input and output signals? __________

3. A_V = __________

4. Draw the output signal.

0 V --

PART TWO QUESTIONS

1. All voltages with respect to ground. No signal applied.

   $V_{B(Q1)}$ = __________, V_{TP1} = __________, $V_{B(Q2)}$ = __________

   $V_{E(Q1,Q2)}$ = __________, V_{RL} = __________

2. Voltage across each transistor: $V_{CE(Q1)}$ = __________, $V_{CE(Q2)}$ = __________

3. Quiescent current with no signal applied: $I_{C(Q)}$ = __________

4. Quiescent power dissipation of *each* transistor: P_D = __________

5. $\mathcal{V}_{RL(\text{max unclipped})}$ = __________ $V_{P\text{-}P}$

6. Collector current with output at maximum signal level: $I_{C(max)}$ = __________

7. Power dissipation of each transistor with maximum output signal: $P_{D(max)}$ = __________
   Show work here.

8. Power dissipated in the load with maximum output signal: $P_{RL(max)}$ = __________
   Show work here.

9. Amplifier efficiency: %Eff = __________
Show work here.

10. Describe what is observed at the output with D_1 then D_2 then both shorted.

11. Explain what causes crossover distortion.

12. Draw your output signal showing the crossover distortion.

0 V --

NOTES AND CALCULATIONS

Class C Amplifiers

EXPERIMENT 17

PURPOSE

The purpose of this experiment is to familiarize you with class C amplifiers. You will build and test an untuned and a tuned class C amplifier.

FORMAL OBJECTIVES

After completing this experiment, you will be able to:

- describe the operation of the class C amplifier in terms of efficiency and duty cycle.
- use the oscilloscope, function generator, and DVM to test a class C amplifier.

PREPARATIONS

In preparation for this experiment, you should study Section 7-3 of *Exploring Electronic Devices*.

COMPONENTS and EQUIPMENT

1 - DC source (12 V, 10 mA capability)
1 - Plug-in experiment board (breadboard)
1 - Function generator with leads
1 - 1 mH High-Q inductor
2 - 10 kΩ, 1/4 W Resistors
1 - 0.01 μF Film capacitor
1 - Digital multimeter with clip leads
1 - Dual-trace oscilloscope with leads
1 - 2N3904 NPN Transistor
1 - 100 kΩ, 1/4 W Resistor
1 - 0.1 μF Film capacitor
1 - 0.001 μF Film capacitor

PART ONE: THE UNTUNED CLASS C COMMON-EMITTER AMPLIFIER

In this part of the experiment, you will assemble and test the class C amplifier shown in Figure 1. You will use the dual-trace oscilloscope to observe and compare the input and output signals of the amplifier. While doing so, you will substitute various values of capacitance for C_C and resistance for R_B to determine the effect on operating duty cycle.

1. Assemble the class C amplifier shown in Figure 1. To start, let $C_C = 0.001\ \mu F$ and $R_B = 10\ k\Omega$.

2. Connect Channel 1 of the oscilloscope to TP1 and Channel 2 to TP3. Set the input mode select for DC on both channels. Trigger from Channel 1.

3. Connect the function generator to the amplifier input and ground.

4. Set the function generator for a frequency of 50 kHz. Select the sine wave function. Set the amplitude control so the sine wave is 2 V_{P-P} at TP1.

5. Move the Channel 1 probe from TP1 to TP2.

6. Apply +12 V to your class C amplifier.

Figure 1

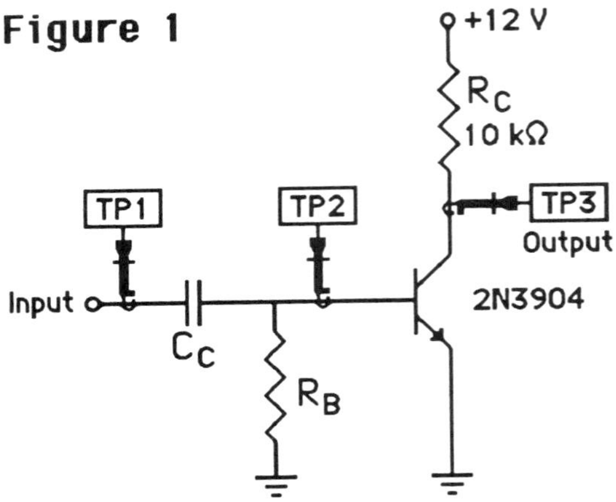

7. Adjust Channel 1 and Channel 2 oscilloscope controls to display three or four cycles of the TP2 signal in the upper half of the screen and TP3 in the lower half.

   Suggested Settings:
   Time/Div Control: 5 μs/div
   Volts/Div Control: Channel 1 = 1 V/div, Channel 2 = 5 V/div

   IMPORTANT! MAKE SURE YOU ARE AWARE OF THE 0-V REFERENCE LINE FOR EACH SIGNAL. MOMENTARILY SET THE INPUT MODE SELECT SWITCHES TO THE GND POSITION TO ESTABLISH THE LOCATION OF THE 0-V REFERENCE LINE.

8. For each combination of C_C and R_B shown on the answer sheet draw the TP2 and TP3 waveforms and determine duty cycle. Clearly mark the 0-V reference line for each signal. Be sure to refer to Section 7-3 of *Exploring Electronic Devices* as an aid to determining duty cycle and conduction angle.

9. Remove power and answer any remaining questions for Part One.

PART TWO: THE TUNED CLASS C COMMON-EMITTER AMPLIFIER

In this part of the experiment, you will replace R_C of your class C amplifier with an LC circuit that is resonant at or near 50 kHz. You will observe the value of having such a circuit.

Figure 2

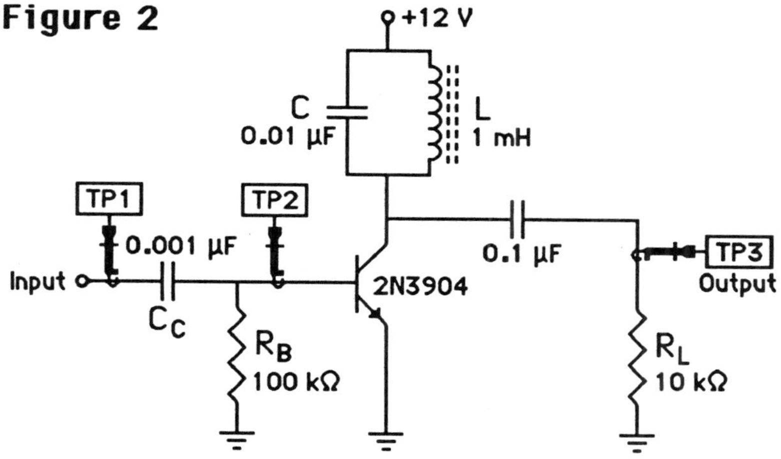

1. Assemble your class C amplifier as shown in Figure 2. Let $C_C = 0.001\ \mu F$ and $R_B = 100\ k\Omega$. Have someone check your work.
2. Follow Steps 2 through 7 of Part One again.
3. Slowly vary the frequency control on the function generator a little bit above and below 50 kHz to obtain the largest peak-to-peak voltage at the amplifier's output at TP3. What you are doing is matching the generator frequency to the actual resonant frequency of the tuned circuit. The calculated resonant frequency is:

   $f_r = 1/(2\pi\sqrt{LC}) = 1 / (2\pi \cdot \sqrt{(1E\text{-}3 \cdot 0.01E\text{-}6)}) = 1 / (2\pi \cdot 3.162E\text{-}6) = 50.3$ kHz

   Note: If a definite peak in output voltage cannot be found, decrease the amplitude of the input signal because the amplifier is being overdriven.
4. Adjust the amplitude of the input signal (signal generator amplitude control) for a maximum undistorted output signal at TP3.
5. Determine the voltage gain of your tuned class C amplifier ($A_V = V_{out}/V_{in} = V_{TP3}/V_{TP1}$) Record this on your answer sheet.
6. Remove power and answer any remaining questions for Part Two.

NAME:______________________________ CLASS:__________ DATE:__________

PART ONE QUESTIONS

1. Draw the TP2 and TP3 waveforms for $C_C = 0.001\ \mu F$ and $R_B = 10\ k\Omega$.

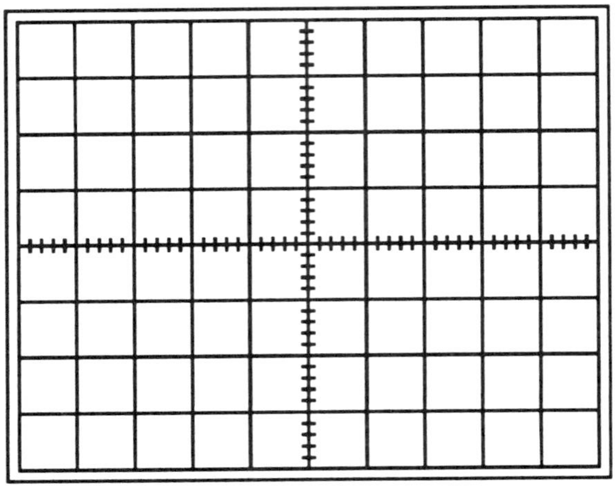

2. T = time for one complete cycle
   = ____________ μs

3. t = transistor conduction time
   = ____________ μs

4. Conduction angle = 360° • t/T
   = ______________________________
   = ____________ degrees

5. % Duty Cycle = 100% • t/T
   = ______________________________
   = ____________%

6. Draw the TP2 and TP3 waveforms for $C_C = 0.1\ \mu F$ and $R_B = 10\ k\Omega$.

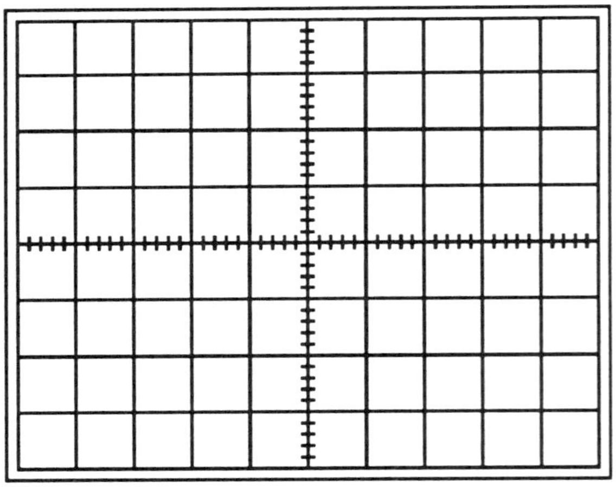

7. t = transistor conduction time
   = ____________ μs

8. Conduction angle = 360° • t/T
   = ______________________________
   = ____________ degrees

9. % Duty Cycle = 100% • t/T
   = ______________________________
   = ____________%

10. Draw the TP2 and TP3 waveforms for $C_C = 0.001\ \mu F$ and $R_B = 100\ k\Omega$.

11. t = transistor conduction time
= ____________ μs

12. Conduction angle = 360° • t/T
= ______________________________
= ____________ degrees

13. % Duty Cycle = 100% • t/T
= ______________________________
= ____________%

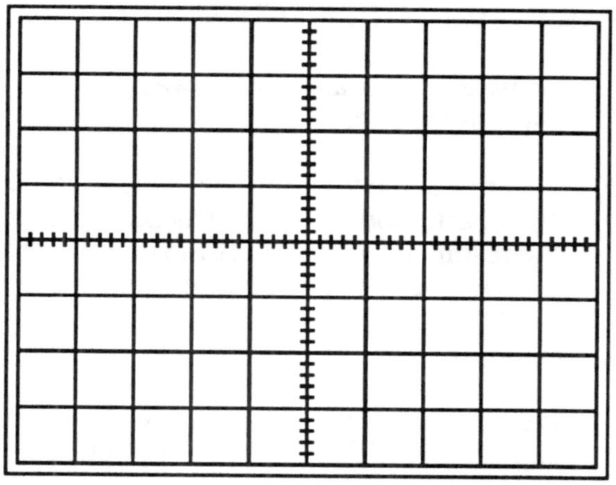

14. Draw the TP2 and TP3 waveforms for $C_C = 0.1\ \mu F$ and $R_B = 100\ k\Omega$.

15. t = transistor conduction time
= ____________ μs

16. Conduction angle = 360° • t/T
= ______________________________
= ____________ degrees

17. % Duty Cycle = 100% • t/T
= ______________________________
= ____________%

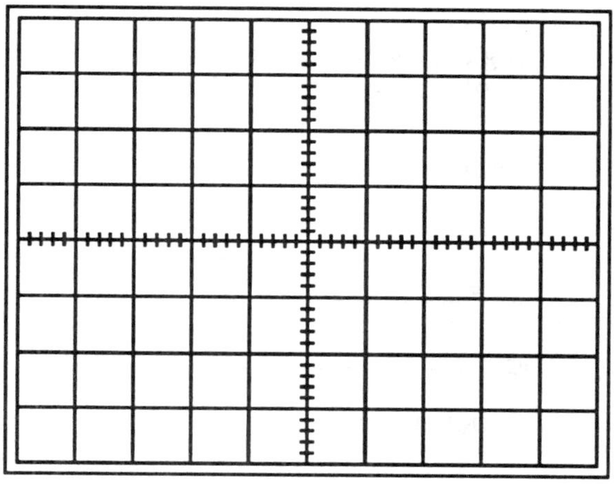

18. Consider the different operating duty cycles and conduction angles resulting from different combinations of C_C and R_B. Write a conclusion that relates capacitor and resistor values to duty cycle and conduction angle.

____
____
____
____

PART TWO QUESTIONS

1. Draw the input and output waveforms on this oscilloscope screen.

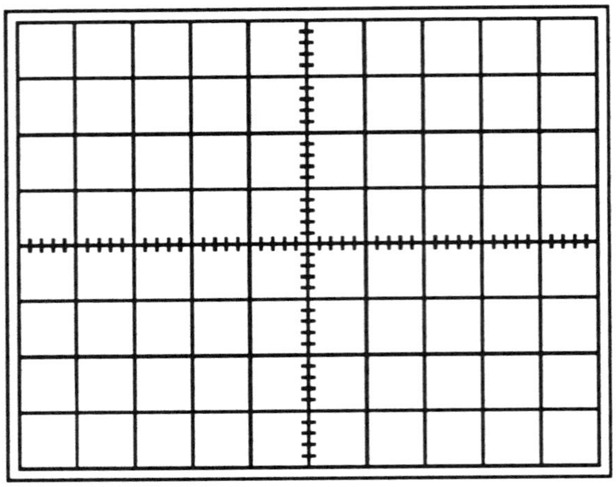

2. Determine the time for one cycle of the output signal after the generator has been matched to the resonant frequency of the tuned circuit:
   T = (no. of div.) • (Time/Div Control)
   = ____________________
   = ____________ μs

3. Determine the actual frequency of the output signal: f = 1/T = ____________ kHz.

4. Determine the voltage gain.
   A_V = ____________________ = ________

5. Using the maximum peak-to-peak output voltage, determine the following: (See Example 7-5 in *Exploring Electronic Devices*.)

   $P_{L(max)}$ = ________________________________

   P_D = ________________________________

   P_{tank} = ________________________________

   P_{DC} = ________________________________

   % Efficiency = ________________________________

NOTES AND CALCULATIONS

Class D Amplifiers

EXPERIMENT 18

PURPOSE

The purpose of this experiment is to familiarize you with class D amplifiers. You will build and test a class D amplifier.

FORMAL OBJECTIVES

After completing this experiment, you will be able to:

- describe the operation of the class D amplifier in terms of efficiency.
- use the oscilloscope, function generator, and DMM to test a class D amplifier.

PREPARATIONS

In preparation for this experiment, you should study Section 7-4 of *Exploring Electronic Devices*.

COMPONENTS and EQUIPMENT

1 - DC source (12 V, 20 mA capability)
1 - Plug-in experiment board (breadboard)
1 - Function generator with leads
1 - 10 kΩ, 1/4 W Resistor
1 - Digital multimeter with clip leads
1 - Dual-trace oscilloscope with leads
1 - 2N3904 NPN Transistor
1 - 1 kΩ, 1/4 W Resistor

PART ONE: THE CLASS D COMMON-EMITTER AMPLIFIER

In this part of the experiment, you will assemble and test the class D amplifier shown in Figure 1. Recall that the class D amplifier is a nonlinear amplifier that operates in either cutoff or saturation, no in-between. You will use your DMM to measure collector current and collector-emitter voltage when the transistor is cutoff and saturated. You will use this information to determine transistor power dissipation and amplifier efficiency.

1. Assemble the class D amplifier of Figure 1.
2. Apply a V_{CC} of +12 V to R_C.
3. Ground R_B and measure I_C and V_{CE}. Record your readings on the answer sheet.
4. Apply +12 V to R_B and measure I_C and V_{CE}. Record your readings on the answer sheet.
5. Remove power and answer all questions for Part One on the answer sheet.

Figure 1

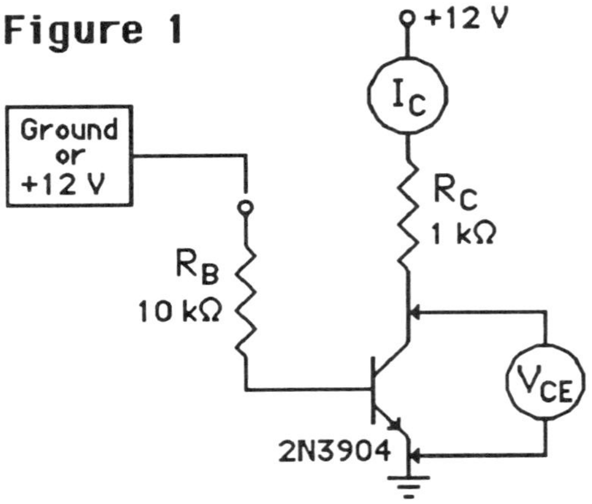

PART TWO: CLASS D AMPLIFICATION

In this part of the experiment, you will apply different waveforms of various amplitudes to the input of the class D amplifier and use the oscilloscope to compare input and output waveforms.

1. Connect the Channel 1 and Channel 2 oscilloscope inputs to the class D amplifier's input and output so you can compare input and output signals.
2. Use your function generator to apply different 1 kHz waveforms to the input of your class D amplifier (sine wave, triangle wave, square wave). The input, of course, is the lose end of R_B.
3. Vary the amplitude of each applied waveform to see how this effects the output waveform. Describe what you see on the answer sheet.
4. Answer all questions on your answer sheet for Part Two.

NAME:______________________________ CLASS:__________ DATE:__________

PART ONE QUESTIONS

1. With R_B grounded: I_C = __________ mA, V_{CE} = __________ V

2. With +12 V applied to R_B: I_C = __________ mA, V_{CE} = __________ V

3. Which condition above causes the transistor to be saturated? __________________

4. Transistor power dissipation when in cutoff = $P_{D(cutoff)}$ = __________________

5. Transistor power dissipation when in saturation = $P_{D(sat)}$ = __________________

6. Total DC power dissipated when transistor is saturated = P_{DC} = __________________

7. Amplifier power efficiency when saturated = % Eff = 100% • (P_{DC} - $P_{D(sat)}$) / P_{DC}
   = ______________________ = __________%

8. When the base voltage is high (V_{BB} = +12 V) the collector voltage is __________ and the transistor is in __________ .

9. When the base voltage is low (V_{BB} = 0 V) the collector voltage is __________ and the transistor is in __________ .

10. How does the efficiency of the class D amplifier compare to other classes of amplifiers?

PART TWO QUESTIONS

1. Describe how the output waveform looked for different input waveforms.

____
____
____
____
____

2. Describe how changing the amplitude of the input waveforms affected the output waveform.

____
____
____
____
____

NOTES AND CALCULATIONS

A Multistage Amplifier

EXPERIMENT 19

PURPOSE

The purpose of this experiment is to familiarize you with a multistage amplifier composed of cascaded transistor stages. You will examine both capacitor and direct coupling of stages.

FORMAL OBJECTIVES

After completing this experiment, you will be able to:

- explain the reason for capacitor coupling.
- predict overall amplifier gain from the schematic of a multistage amplifier.
- use the function generator and oscilloscope to test a multistage amplifier.

PREPARATIONS

In preparation for this experiment, you should review Chapters 6 and 7 of *Exploring Electronic Devices.*

COMPONENTS and EQUIPMENT

1 - Analog or digital multimeter
1 - Function generator with test leads
2 - 2N3904 NPN Transistors
2 - 1 μF, 25 V Electrolytic capacitors
1 - 10 kΩ, 1/4 W Resistor
1 - 1.5 kΩ, 1/4 W Resistor
1 - 330 Ω, 1/4 W Resistor
1 - 51 Ω, 1/4 W Resistor
1 - Plug-in experiment board (breadboard)
1 - Dual-trace oscilloscope with probes
1 - 2N2219 NPN Transistor
1 - 30 kΩ, 1/4 W Resistor
1 - 3.3 kΩ, 1/4 W Resistor
1 - 680 Ω, 1/4 W Resistor
2 - 100 Ω, 1/4 W Resistors

DISCUSSION

In the following steps, you will construct and test the multistage amplifier of Figure 1. Before construction and testing begin, you will use your circuit analysis skills to predict quiescent DC voltages and amplifier gain. Then you will build the circuit and make measurements to verify your calculations. As you examine Figure 1, you will immediately notice that stages 1 and 2 are capacitively coupled and stages 2 and 3 are directly coupled. When direct coupling is used, the DC output voltage of the previous stage is used to bias the succeeding stage. Thus, the AC signal is passed on a DC reference. When capacitor coupling is used, the DC output voltage of the previous stage is blocked from getting to the next stage. Capacitor coupling is typically used to couple cascaded common-emitter amplifiers because the DC output voltage (collector voltage) of the first stage is usually much higher than the DC bias voltage needed by the second.

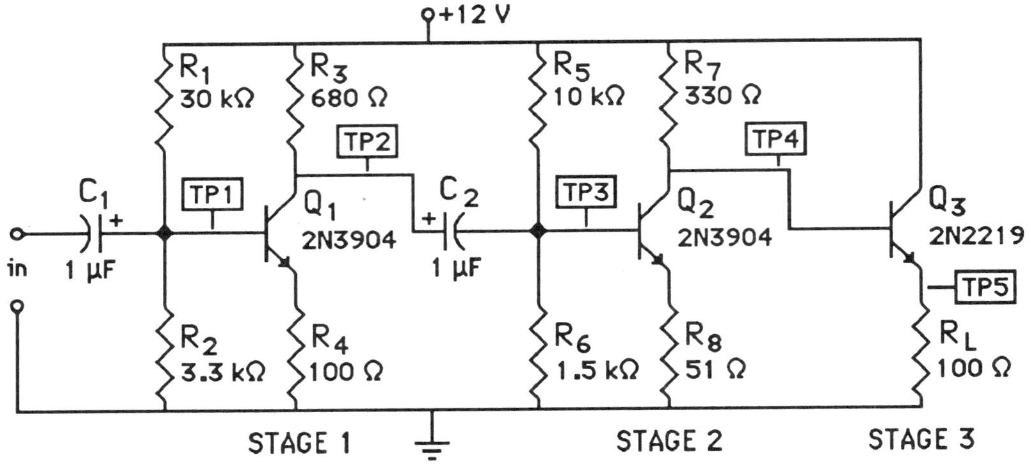

Figure 1

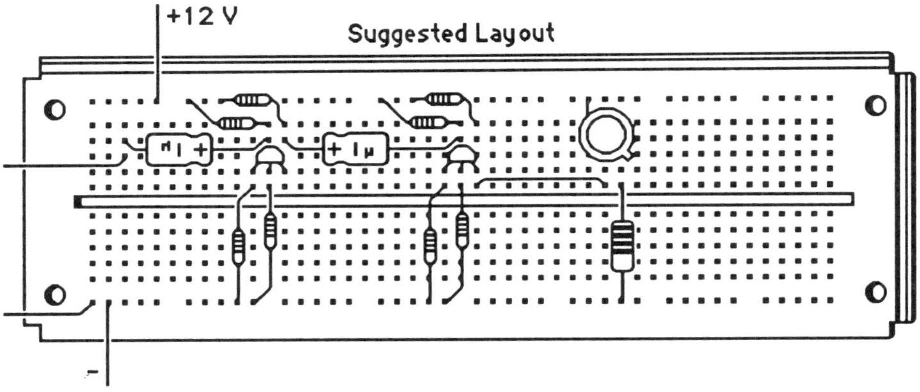

PROCEDURE

1. Take time to perform a thorough mathematical, or theoretical, analysis on the circuit of Figure 1. Calculate the DC voltages at all test points. Calculate the AC voltages at all test points and assume that a 100 $mV_{P\text{-}P}$ test signal is applied to the input of the amplifier. Record all of your calculations on the answer sheet.

Note: You will need to calculate the amplification factor for each stage. You should be aware that the base impedance of Q_3 is in parallel with R_7 and should be used in calculating A_V for Q_2. Likewise, the input impedance to stage two is in parallel with R_3 and must be used in calculating A_V for Q_1. As a realistic estimate, assume the ß of each transistor to be 150 for your calculations.

2. Construct the circuit shown in Figure 1. Apply Power.

   Note: If the amplifier is unstable, place a 100 pF capacitor from the collector to the base of Q_2.

3. Without any AC test signal applied, measure and record the DC voltage at each testpoint.

4. Compare these measurements with your predicted values. If there is a great difference, try to determine what you did wrong. Ask your instructor for assistance if you cannot determine the problem.

5. Apply a 100 mV_{P-P}, 1 kHz test signal to the input of your amplifier. Use your oscilloscope to measure the signal voltage at each testpoint. Record your findings on the answer sheet.

6. How do these values compare to your predicted ones?

7. Use the measured values of signal voltage to determine the amplification factor for each stage and record them on the answer sheet.

8. Remove power and turn off your equipment. Answer any remaining questions on the answer sheet.

NAME:______________________ CLASS:__________ DATE:__________

QUESTIONS

1. Calculate the DC voltages at all testpoints marked in Figure 1.

   V_{TP1} = ______________________, V_{TP2} = ______________________

   V_{TP3} = ______________________, V_{TP4} = ______________________

   V_{TP5} = ______________________

2. Z_{in} for Q_3 = ______________________

   Z_{in} for Q_2 = ______________________

3. Determine A_V for each stage.

   A_V stage 1 = __________, A_V stage 2 = __________, A_V stage 3 = __________

4. Assume a 100 $mV_{P\text{-}P}$ AC signal is applied to the input of the amplifier. Predict the AC voltage at each testpoint.

   v_{TP1} = ______________________, v_{TP2} = ______________________

   v_{TP3} = ______________________, v_{TP4} = ______________________

   v_{TP5} = ______________________

5. What configuration is the stage 3 amplifier? (a) CE (b) CB (c) CC

6. What is the primary purpose for the third stage of the amplifier?

   ____

7. Record the measured DC voltages for all testpoints of Figure 1.

V_{TP1} = __________, V_{TP2} = __________, V_{TP3} = __________

V_{TP4} = __________, V_{TP5} = __________

8. Record the measured AC signal voltages for all testpoints of Figure 1.

$\mathcal{V}_{TP1}$ = __________, $\mathcal{V}_{TP2}$ = __________, $\mathcal{V}_{TP3}$ = __________

$\mathcal{V}_{TP4}$ = __________, $\mathcal{V}_{TP5}$ = __________

9. Use the measured AC signal voltages to determine the amplification factor for each stage.

A_V stage 1 = ____________________

A_V stage 2 = ____________________

A_V stage 3 = ____________________

10. Determine the gain in decibels for each stage. ($\#dB = 20 \log(A_V)$)

Stage 1 gain = ____________________

Stage 2 gain = ____________________

Stage 3 gain = ____________________

11. Determine the overall amplifier gain in decibels.
($\#dB = 20 \log(A_{V(total)})$ and $A_{Vtotal} = A_{V1} \bullet A_{V2} \bullet A_{V3}$)
Also, $\#dB \text{ overall} = dB_1 + dB_2 + dB_3$

#dB overall = ____________________

12. Why is C_2 placed in the circuit with its positive plate toward the collector of Q_1?

13. What would happen if C_2 became shorted?

14. What would happen if Q_1 became shorted from collector to base?

NOTES AND CALCULATIONS

JFET Operating Characteristics

EXPERIMENT

20

PURPOSE

The purpose of this experiment is to familiarize you with the operating characteristics of JFET amplifiers. You will test a common JFET and develop its transconductance curve and a family of characteristic curves.

FORMAL OBJECTIVES

After completing this experiment, you will be able to:

- describe the relationship between gate-to-source voltage and drain current for a JFET.
- use the DMM to determine operating characteristics for a JFET.

PREPARATIONS

In preparation for this experiment, you should study Sections 8-1 and 8-2 of *Exploring Electronic Devices.*

COMPONENTS and EQUIPMENT

2 - DC Sources (-9 V and +15 V)
1 - Plug-in experiment board (breadboard)
3 - 10 kΩ Trim-pots
1 - Digital multimeter with clip leads
1 - MPF102 n-channel JFET
1 - 10 kΩ, 1/4 W Resistor

PART ONE: THE TRANSCONDUCTANCE CURVE

In this part of the experiment, you will assemble the test circuit of Figure 1 and use it to determine the actual transconductance curve, also known as the transfer curve, for your n-channel JFET. Recall that the transfer curve is a graph of the relationship between gate-to-source voltage and drain current. As a transconductance curve, it graphically shows the relationship between a change in gate-to-source voltage and a resulting change in drain current where $g_m = \Delta I_D / \Delta V_{GS}$. Take time to review Section 8-2 in *Exploring Electronic Devices.*

Figure 1

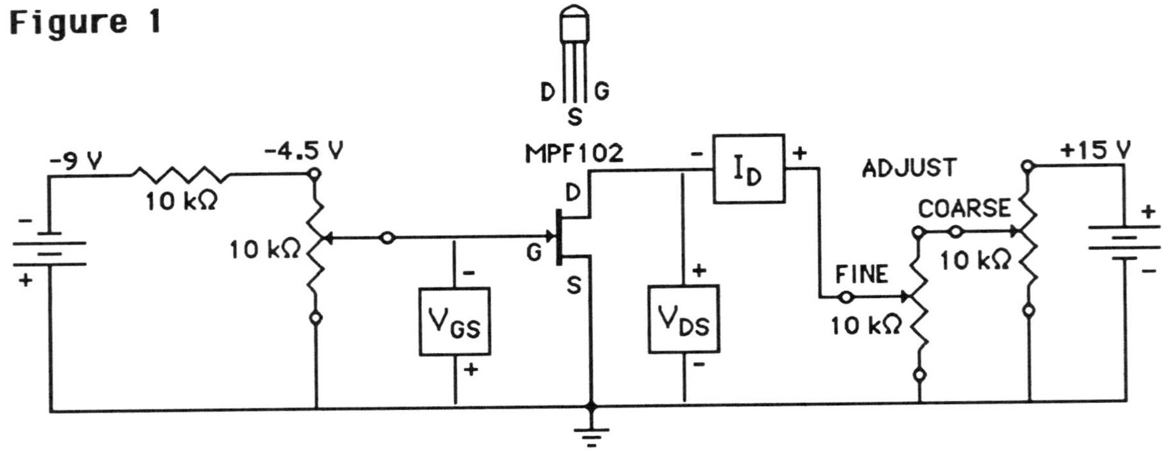

1. Carefully assemble the test circuit of Figure 1. Ask someone to check your wiring before you apply power.
2. Set the drain potentiometers for maximum drain voltage (+15 V).
3. Set the gate potentiometer for maximum negative voltage (-4.5 V).
4. In this procedure, you will decrease the gate-to-source voltage a half a volt at a time and measure the drain current. Record the drain current for each gate-to-source voltage setting on the answer sheet for Part One.

   Note: It is better to use two multimeters for this task - one to measure drain current and the other to measure gate-to-source voltage. However, one multimeter can be used to accomplish the task by moving it back and forth, switching it from voltmeter to ammeter.
5. Remove power.
6. Use your collected data to plot and draw the transfer curve for your JFET on Chart 1.
7. Clearly mark $V_{GS(off)}$ and I_{DSS} on the chart.
8. Answer any other questions for Part One on the answer sheet.

PART TWO: A FAMILY OF CHARACTERISTIC CURVES

In this part of the experiment, you will use the test circuit of Figure 1 to develop a family of characteristic curves for your JFET. Recall that the JFET will act as a constant-current source when saturated. The amount of saturation current is dependant upon the gate-to-source bias voltage. You will demonstrate this to yourself in this part of the experiment.

1. Apply power to your circuit. Set the drain potentiometers to 0 V. Set the gate potentiometer to -4.0 V.

2. In this part, you will set the gate potentiometer in one-volt steps. For each step-setting of the gate potentiometer, you will increase the drain voltage a half volt at a time and record the resulting drain current. The gate voltage shall not be changed until you have recorded the drain current for all drain-to-source settings from 0 V to +15 V. All currents shall be recorded in Table 1 on the answer sheet.

   Note: As before, it is best to use two multimeters - one to measure gate-to-source and drain-to-source voltages and one to measure drain current.

3. When Table 1 is complete, remove power and plot and draw the drain curves for each setting of the gate potentiometer. Plot your family of drain curves on Chart 2.

4. Answer any other questions for Part Two on the answer sheet.

NOTES AND CALCULATIONS

NAME:________________________________ CLASS:__________ DATE:__________

PART ONE QUESTIONS

1. For $V_{GS} = -4.0$ V, I_D = __________ mA.
2. For $V_{GS} = -3.5$ V, I_D = __________ mA.
3. For $V_{GS} = -3.0$ V, I_D = __________ mA.
4. For $V_{GS} = -2.5$ V, I_D = __________ mA.
5. For $V_{GS} = -2.0$ V, I_D = __________ mA.
6. For $V_{GS} = -1.5$ V, I_D = __________ mA.
7. For $V_{GS} = -1.0$ V, I_D = __________ mA.
8. For $V_{GS} = -0.5$ V, I_D = __________ mA.
9. For $V_{GS} = 0.0$ V, I_D = __________ mA.

Chart 1

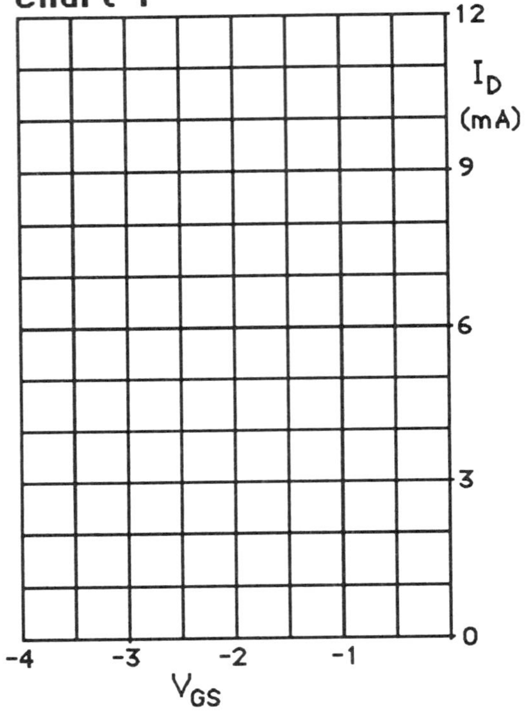

10. $V_{GS(off)}$ = __________ V, I_{DSS} = __________ mA

Optional Exercise: Use Design Note 8-1 to develop this transfer curve and see how they compare.

PART TWO QUESTIONS

TABLE 1

V_{GS} =	-3 V	-2 V	-1 V	0 V
V_{DS}	I_D (mA)	I_D (mA)	I_D (mA)	I_D (mA)
0.0 V				
0.5 V				
1.0 V				
1.5 V				
2.0 V				
2.5 V				
3.0 V				
3.5 V				
4.0 V				
6.0 V				
8.0 V				
10.0 V				
12.0 V				
15.0 V				

Plot each column of drain-current data on Chart 2.

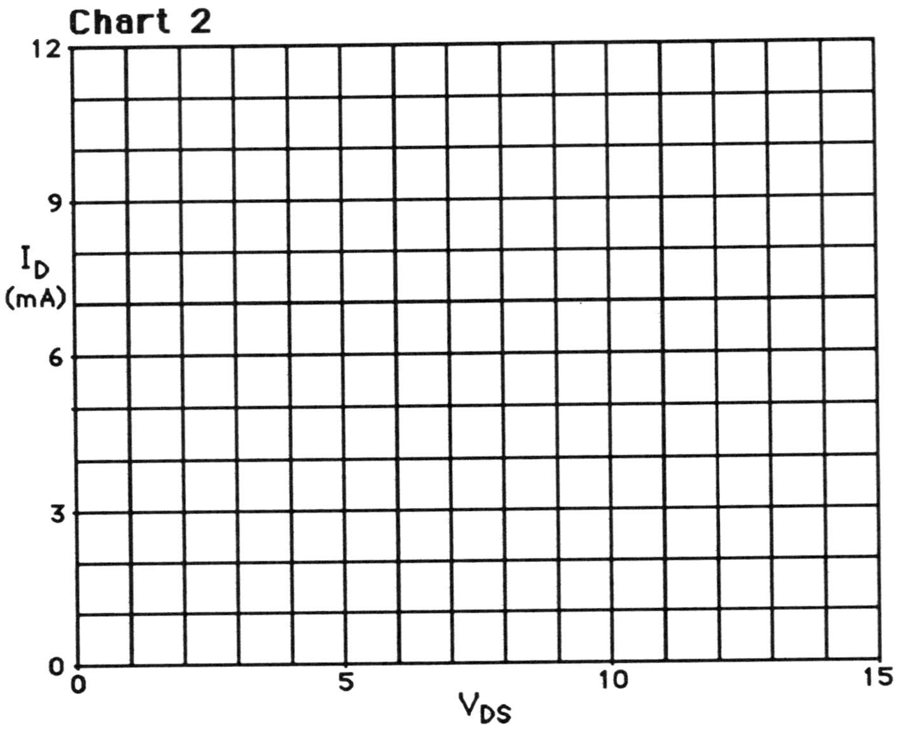

1. Which of the drain curves represents I_{DSS}? ______________________

2. Describe how V_{GS} affects the saturation current of the JFET.

____

____

3. What affect does a large change in V_{DS} have on drain current? Assume the large change is from +1 V to +15 V.

____

____

4. Explain why the JFET makes a fairly good constant-current source.

____

____

Note: You will use your family of drain curves in the next experiment.

JFET Biasing

EXPERIMENT 21

PURPOSE

The purpose of this experiment is to familiarize you with JFET biasing. You will assemble a common-source amplifier and investigate gate biasing and self biasing.

FORMAL OBJECTIVES

After completing this experiment, you will be able to:

- describe the effect on JFET amplifier gain when gate bias is varied.
- self-bias a common-source JFET amplifier.
- use the multimeter and oscilloscope to analyze a JFET amplifier.

PREPARATIONS

In preparation for this experiment, you should study Section 8-3 of *Exploring Electronic Devices*.

COMPONENTS and EQUIPMENT

2 - DC sources (-9 V and +15 V)
1 - Plug-in experiment board (breadboard)
1 - Function generator with leads
1 - 10 kΩ Trim-pot
1 - 100 kΩ, 1/4 W Resistor
1 - 1 kΩ, 1/4 W Resistor
1 - Multimeter with clip leads
1 - Dual-trace oscilloscope with leads
1 - MPF102 n-channel JFET
1 - 0.1 μF Capacitor
1 - 27 kΩ, 1/4 W Resistor
1 - 150 Ω, 1/4 W Resistor

PART ONE: BIAS VOLTAGE AND AMPLIFIER GAIN

In this part of the experiment, you will assemble and test the common-source JFET amplifier shown in Figure 1. The potentiometer will allow you to vary the gate bias voltage over a range of approximately -2.4 V to 0 V. At various settings of bias voltage, you will measure DC parameters and determine AC gain using an applied test signal at the input. You will then draw conclusions regarding the relationship between gate bias voltage and amplification factor.

Figure 1

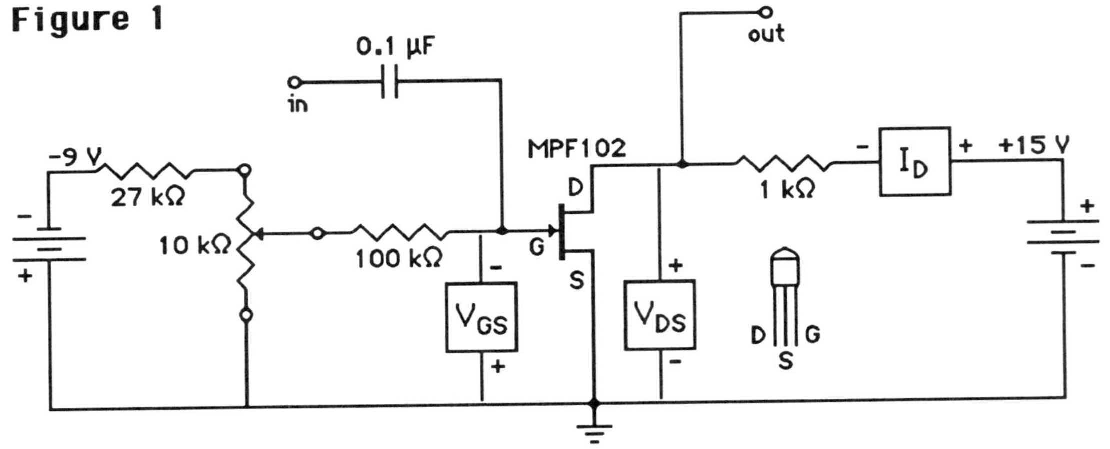

1. Construct the circuit shown in Figure 1. Use the MPF102 that you used in Experiment 20.
2. Set V_{GS} to -2 V by adjusting the potentiometer. Measure and record V_{DS} and I_D on your answer sheet.
3. Connect Channel 1 of the oscilloscope to the gate (input) of your JFET. Connect Channel 2 to the drain (output). Set the input mode select for AC on each channel.
4. Apply a 1 kHz sine wave test signal to the input of the amplifier. Adjust the amplitude of the input signal to 100 $mV_{P\text{-}P}$. Adjust the oscilloscope controls so the input signal is in the top half of the screen.
5. Observe the output signal on Channel 2 of the oscilloscope. Adjust the oscilloscope controls so the output signal is in the lower half of the screen. Measure the peak-to-peak voltage of the output signal and record it on the answer sheet for Part One.
6. Determine the amplification factor for your amplifier and convert it to gain in decibels. Record your answers on the answer sheet.
7. Repeat the above procedure with V_{GS} set at -1 and 0 V.
8. Remove power. Answer any remaining questions for Part One.

PART TWO: THE SELF-BIASED COMMON-SOURCE JFET AMPLIFIER

In this part of the experiment, you will assemble the circuit shown in Figure 2 and use it to demonstrate self bias.

Figure 2

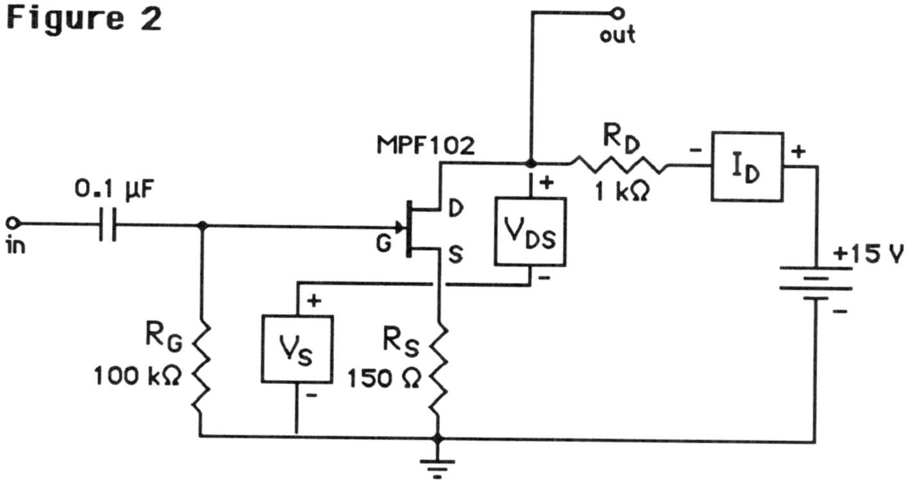

1. Construct the circuit shown in Figure 2. Use the same MPF102.

2. With no test signal applied, measure and record V_G, V_{GS}, V_S, V_{DS} and I_D on your answer sheet.

3. Use Chart 2 from Experiment 20 and draw the DC load line. You will have to project $I_{D(sat)}$ above the top of the chart since it is near 13 mA and the chart stops at 12 mA.

4. From your measurements in Step 2, mark the quiescent operating point on the load line. Mark $I_{D(Q)}$ and $V_{DS(Q)}$.

5. Connect Channel 1 of the oscilloscope to the gate (input) of your JFET. Connect Channel 2 to the drain (output). Set the input mode select for AC on each channel.

6. Apply a 1 kHz sine wave test signal to the input of the amplifier. Adjust the amplitude of the input signal to 100 $mV_{P\text{-}P}$. Adjust the oscilloscope controls so the input signal is in the top half of the screen.

7. Observe the output signal on Channel 2 of the oscilloscope. Adjust the oscilloscope controls so the output signal is in the lower half of the screen. Measure the peak-to-peak voltage of the output signal and record it on the answer sheet for Part Two.

6. Determine the amplification factor for your amplifier and convert it to gain in decibels. Record your answers on the answer sheet.

7. Increase the amplitude of the input signal until distortion is observed in the output signal. Distortion is any change in the symmetrical sinusoidal shape of the waveform. Record the maximum undistorted peak-to-peak output voltage on the answer sheet.

8. Mark the maximum undistorted peak-to-peak output voltage on the baseline of Chart 2.

9. Remove power. Answer all remaining questions for Part Two.

NAME:____________________ CLASS:__________ DATE:__________

PART ONE QUESTIONS

1. For $V_{GS} = -2$ V: $V_{DS} =$ __________ V, $I_D =$ __________ mA

   $v_{out} =$ __________ $V_{P\text{-}P}$, $A_V =$ __________, Gain in dB = __________ dB

2. For $V_{GS} = -1$ V: $V_{DS} =$ __________ V, $I_D =$ __________ mA

   $v_{out} =$ __________ $V_{P\text{-}P}$, $A_V =$ __________, Gain in dB = __________ dB

3. For $V_{GS} = 0$ V: $V_{DS} =$ __________ V, $I_D =$ __________ mA

   $v_{out} =$ __________ $V_{P\text{-}P}$, $A_V =$ __________, Gain in dB = __________ dB

4. What is the phase relationship between the input and output signals? __________

5. From the data collected above, how does V_{GS} affect A_V?

   ____

   ____

PART TWO QUESTIONS

1. $V_G =$ __________ V, $V_{GS} =$ __________ V, $V_S =$ __________ V

   $V_{DS} =$ __________ V, $I_D =$ __________ mA

2. $I_{D(sat)} =$ __________ mA, $V_{DS(cutoff)} =$ __________ V

3. $I_{D(Q)} =$ __________ mA, $V_{DS(Q)} =$ __________ V

4. $A_V =$ __________, Gain in dB = __________ dB

5. $v_{out(max)} =$ __________ $V_{P\text{-}P}$

6. If it is desired that V_{GS} be -0.5 V, what value should R_S be? Show calculations.

   ____

NOTES AND CALCULATIONS

E-Type Power MOSFET

EXPERIMENT 22

PURPOSE

The purpose of this experiment is to familiarize you with enhancement type power MOSFETs.

FORMAL OBJECTIVES

After completing this experiment, you will be able to:

- describe the operating characteristics of enhancement type MOSFETs.
- use the oscilloscope, function generator, and multimeter to test an E-type MOSFET.

PREPARATIONS

In preparation for this experiment, you should study Sections 8-5 and 8-6 of *Exploring Electronic Devices*.

Also, it will be necessary to mount your MOSFET to a small clip-on heat sink and solder short #22 wires to the MOSFET's leads so it may be plugged into your protoboard. As an alternative, clip leads may be used, though less desirable.

COMPONENTS and EQUIPMENT

1 - DC source (12 V, 1 A capability)
1 - Plug-in experiment board (breadboard)
1 - Function generator with leads
1 - 10 kΩ, 1/4 W Potentiometer
1 - 1,000 μF, 35 V Capacitor
2 - Multimeters with clip leads
1 - Dual-trace oscilloscope with leads
1 - IRF511 Power MOSFET with heat sink
1 - 470 Ω, 1/4 W Resistor

PART ONE: THE CLASS D COMMON-EMITTER AMPLIFIER

In this part of the experiment, you will assemble and test the power MOSFET circuit shown in Figure 2. The MOSFET used in this experiment is the IRF511 E-type power MOSFET shown in Figure 1. Notice the internal diode from drain to source. This diode serves to protect the MOSFET from high transient voltages generated by collapsing fields of inductive loads such as motors and relays. Also, notice the maximum ratings for this device. In this experiment, you will not even come close to the maximum ratings. However, be aware that the MOSFET and the 20-Ω resistor will get very warm during this first part of the experiment. If the MOSFET becomes too hot to touch, remove power immediately and let it cool. Obtain voltage measurements as quickly as possible.

Figure 1

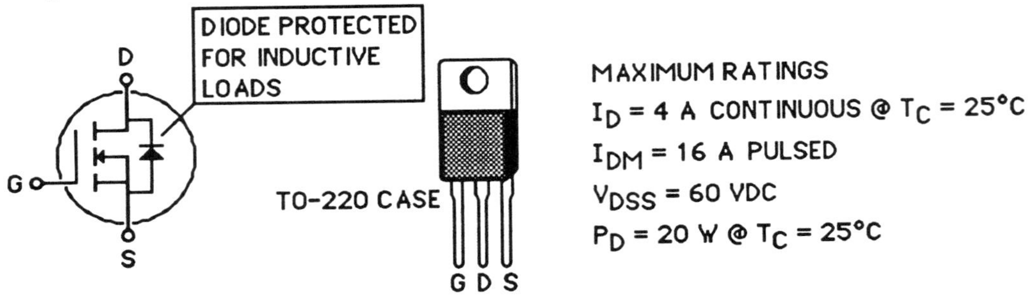

IRF511 - MOTOROLA TMOS E-TYPE POWER MOSFET

1. Assemble the test circuit of Figure 2. Do not apply power until you are ready to take measurements.

Figure 2

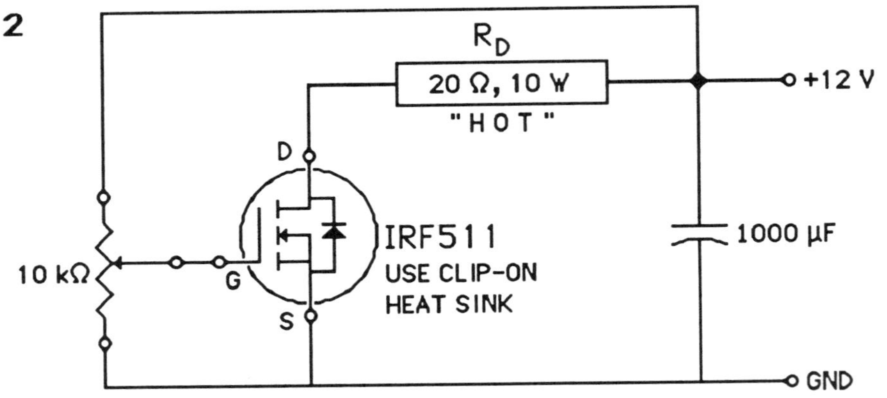

2. It is highly recommended that you use two voltmeters here to read V_{GS} and V_{DS} at the same time. Here, you will set the gate voltage to the settings specified in Table 1 and record the corresponding drain-to-source voltage. Apply power and proceed. *Remove power when finished.*

CAUTION! THE RESISTOR AND MOSFET MAY BECOME HOT!

3. Use the measured drain-to-source voltages to calculate V_{RD} and I_D in Table 1.

   $V_{RD} = +12\ V - V_{DS}$ and $I_D = V_{RD}/20\Omega$

4. Use Chart 1 to make a double graph. Graph I_D as a function of V_{GS} and graph V_{DS} as a function of V_{GS}.

5. Answer any remaining questions for Part One.

PART TWO: THE TUNED CLASS C COMMON-EMITTER AMPLIFIER

In this part of the experiment, you will operate the MOSFET as a class D amplifier which is the usual mode of operation for E-type power MOSFETs. Your circuit will be that of Figure 3.

Figure 3

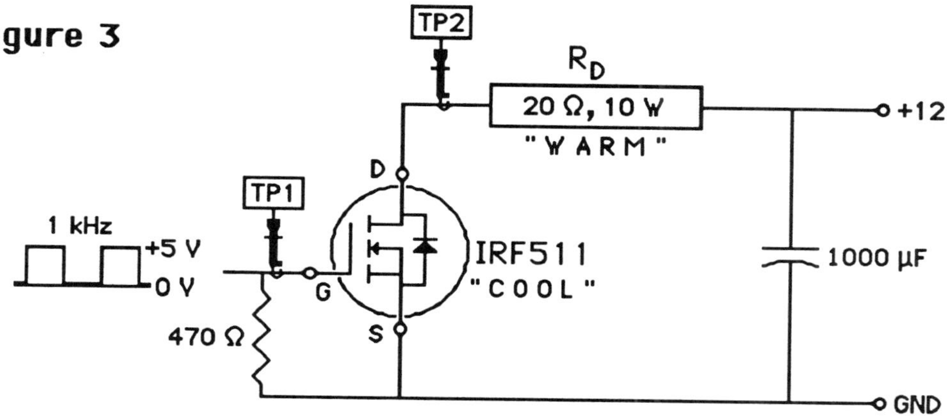

1. Connect your MOSFET as shown in Figure 3.

2. Set the function generator to produce a 1-kHz pulsating square wave having a 50% duty cycle (that means the square wave is on half the time). Connect the function generator to the gate of your MOSFET.

3. Connect Channel 1 of the oscilloscope to the gate (TP1) and Channel 2 to the drain (TP2). Adjust the oscilloscope so input and output signals can be compared. The input mode select switch should be set on DC for each channel.

4. Apply power (+12 V).

5. Answer all questions for Part Two on the answer sheet as you observe the input and output waveforms.

NAME:______________________________ CLASS:__________ DATE:__________

PART ONE QUESTIONS

TABLE 1

V_{GS} (V)	V_{DS} (V)	V_{RD} (V)	I_D (mA)
0.0			
0.5			
1.0			
1.5			
2.0			
2.5			
3.0			
3.5			
4.0			
5.0			
6.0			
8.0			
10.0			
12.0			

1. From the data in Table 1, at what value of V_{GS} does the MOSFET become saturated? ________
2. Use the values of V_{DS} and I_D for V_{GS} = 10 V to determine $r_{DS(on)}$. ________________
3. Calculate the power dissipated by the MOSFET when it is fully saturated.

   ____

4. Calculate the power dissipated by the MOSFET when it is cutoff.

   ____

Chart 1

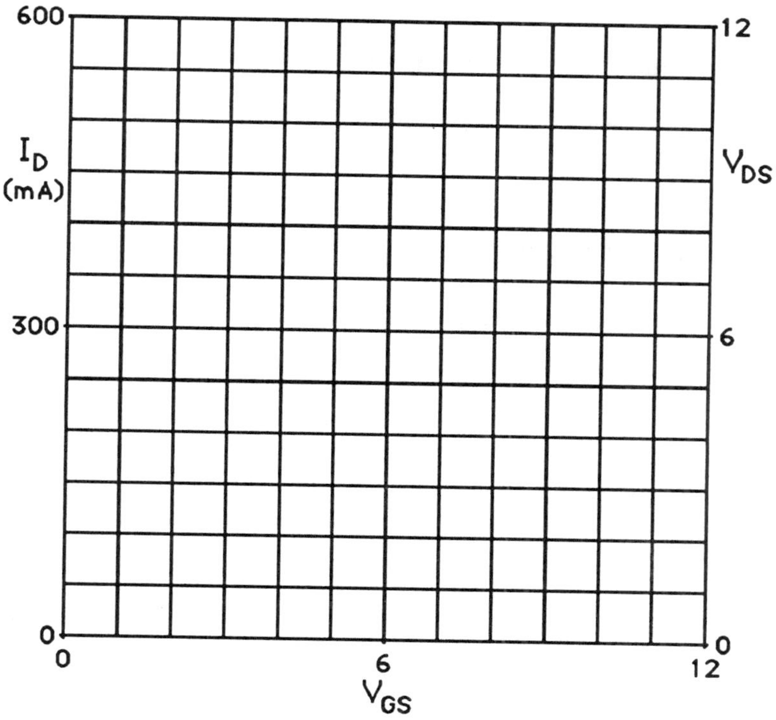

PART TWO QUESTIONS

1. What is the phase relationship between the input and output waveforms? ____________
2. Describe the type of output waveform.

____

____

3. What is the maximum, or peak, output voltage? ____________
4. What is the minimum output voltage? ____________
5. At what points in the output waveform is the MOSFET saturated?

____

6. What is the average voltage across the 20-Ω drain resistor? ____________
7. What is the average current through the 20-Ω drain resistor? ____________
8. Calculate the amount of power dissipated by the drain resistor.

____

9. What is the average amount of voltage across the MOSFET? ____________

10. What is the average amount of current through the MOSFET? ____________

11. Approximately how much power is the MOSFET dissipating?

____

12. Explain why the MOSFET is running cool.

____

____

NOTES AND CALCULATIONS

Amplifier Low-Frequency Response

EXPERIMENT 23

PURPOSE

The purpose of this experiment is to give you the opportunity to see the effect coupling capacitors have on the low-frequency response of amplifiers. You will also learn how to use standard laboratory equipment to determine frequency response.

FORMAL OBJECTIVES

After completing this experiment, you will be able to:

- describe the effect coupling capacitors have on amplifier frequency response.
- use a frequency generator and an oscilloscope to determine actual cutoff frequencies.

PREPARATIONS

In preparation for this experiment, you should study Sections 9-1, 9-2, and 9-5 of *Exploring Electronic Devices.*

COMPONENTS and EQUIPMENT

1 - Dual-trace oscilloscope with probes
1 - Function generator with clip leads
1 - 2N3904 NPN Transistor
1 - 0.01 μF Film capacitor
2 - 10 kΩ, 1/4 W Resistor
1 - 680 Ω, 1/4 W Resistor
1 - Plug-in experiment board (breadboard)
1 - DC Source (12 V, 10 mA capability)
1 - 0.1 μF Film capacitor
1 - 68 kΩ, 1/4 W Resistor
1 - 4.7 kΩ, 1/4 W Resistor

DISCUSSION

In the following steps, you will analyze, assemble, and test the CE amplifier circuit of Figure 1. The focus of this experiment is the lowpass filter characteristics of the input and output coupling networks involving C_1 and C_2. As you should be aware, the capacitive reactance of the input and output coupling capacitors increases as frequency applied to them decreases. This results in an increasing voltage drop across the capacitors and less voltage for the load resistor. The input and output networks, illustrated in Figure 2, form lowpass filters whose slopes of rolloff are ideally 6 dB/octave or 20 dB/decade. Since the signal must pass through both of these networks, the slope of rolloff is further increased, nearly doubled, below the lowest cutoff frequency. In this experiment you will determine the two cutoff frequencies and which is dominant.

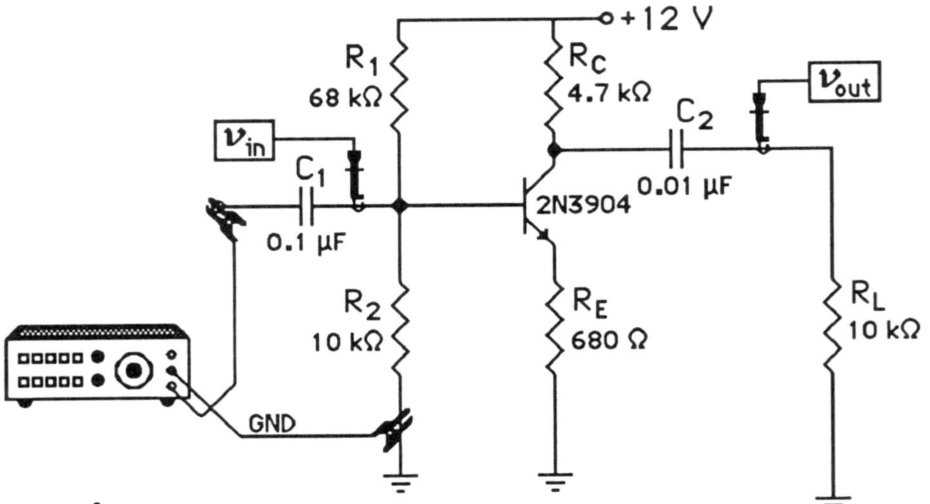

Figure 1

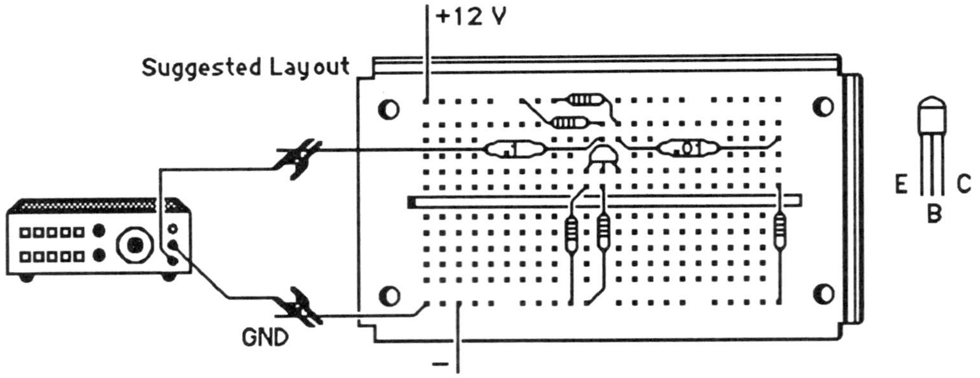

Note: The coupling capacitors are deliberately low-valued in this amplifier to allow you to more easilly identify the cutoff frequencies. If this amplifier was intended to amplify very low audio frequencies, the coupling capacitors would be closer to 10 μF.

Figure 2

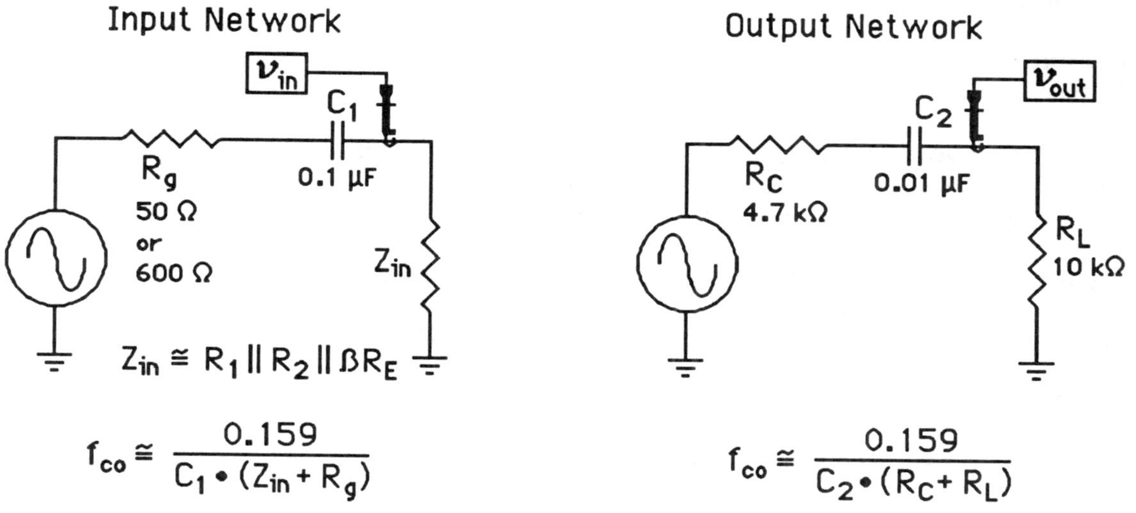

PROCEDURE

1. Take time to perform a mathematical analysis of the low-frequency characteristics of the amplifier in Figure 1. In other words, determine the lowpass cutoff frequency for the input and output coupling networks. Note that for the input network you must determine the input impedance first. Formulas are provided in Figure 2. Show all work on the answer sheet.

2. Construct the circuit shown in Figure 1. Apply power.

3. Set the function generator for sine wave. Adjust the generator's amplitude control for an output signal of 1 $V_{P\text{-}P}$. The frequency of the generator signal will be varied over a range of 10 kHz down to 10 Hz.

4. Connect the oscilloscope to the base of the transistor and measure and record peak-to-peak input voltages for all frequencies listed in Table 1 on the answer sheet. Remove power when finished.

   Note: Take care not to bump the amplitude control on the function generator. You should only adjust the frequency control and range switches.

5. Convert each voltage measurement to decibels and record them in the next column of Table 1. The voltage measured at a frequency of 10 kHz should be approximately 1 $V_{P\text{-}P}$ and shall be considered as 0 dB and will be used as a reference for all other voltages. For example:

   #dB = 20 log (v_{in}/1 $V_{P\text{-}P}$) = 20 log (1 $V_{P\text{-}P}$/1 $V_{P\text{-}P}$) = 20 log (1) = 0 dB

   If the voltage at 5 kHz is 960 m$V_{P\text{-}P}$, then #dB = 20 log (960 m$V_{P\text{-}P}$/1 $V_{P\text{-}P}$) = -0.35 dB

   If the voltage at 1 kHz is 900 m$V_{P\text{-}P}$, then #dB = 20 log (900 m$V_{P\text{-}P}$/1 $V_{P\text{-}P}$) = -0.92 dB

6. Reapply power to your circuit. Adjust the function generator's amplitude and frequency controls so the voltage at the *load resistor* is 1 V_{P-P} at a frequency of 10 kHz.

7. Measure and record peak-to-peak output voltages for all frequencies listed in Table 1. Remove power.

8. Convert each voltage measurement to decibels and record them in the next column of Table 1. The voltage measured at a frequency of 10 kHz should be approximately 1 V_{P-P} and shall be considered as 0 dB and will be used as a reference for all other voltages.

9. Use the input frequency response data (dB data) to plote a graph on Chart 1.

10. Now, use the output frequency response data to plot a second graph on Chart 1.

11. Clearly label each graph on Chart 1 as either input or output.

12. Answer all remaining questions on the answer sheet.

NAME:______________________________ CLASS:__________ DATE:__________

QUESTIONS

1. Mathematically determine the cutoff frequency for the input network of the amplifier. Show all work.

____

____

____

2. Mathematically determine the cutoff frequency for the output network of the amplifier. Show all work.

____

____

____

TABLE 1

f (Hz)	$V_{in(P-P)}$	#dB	$V_{out(P-P)}$	#dB
10 k	______	0	______	0
5 k	______	______	______	______
2 k	______	______	______	______
1 k	______	______	______	______
500	______	______	______	______
200	______	______	______	______
100	______	______	______	______
50	______	______	______	______
10	______	______	______	______

PLOT FREQUENCY RESPONSE GRAPHS ON CHART 1.

Chart 1

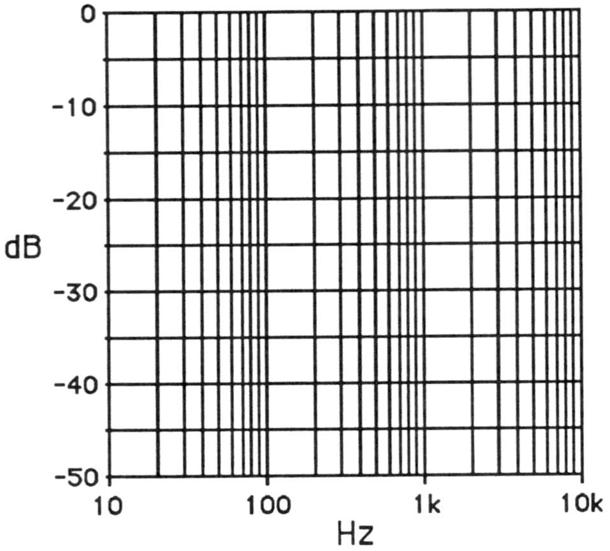

3. From the first graph plotted on Chart 1, what is the approximate cutoff frequency of the input network? (Recall the cutoff frequency is at -3 dB.) $f_{co(input)}$ = __________ Hz

4. From the first graph plotted on Chart 1, what is the slope of rolloff from 100 Hz down to 10 Hz? __________dB/decade

5. From the second graph plotted on Chart 1, what are the two break frequencies and which is dominant? $f_{co(dominant)}$ = __________ Hz, $f_{co(2)}$ = __________ Hz

6. From the second graph plotted on Chart 1, what is the slope of rolloff from:

   (a) 1 kHz down to 100 Hz? __________dB/decade

   (b) 100 Hz down to 10 Hz? __________dB/decade

7. Write a brief description of how you would use the function generator and oscilloscope to determine the actual cutoff frequency of the input network.

____

____

____

____

____

____

____

____

The Differential Amplifier

EXPERIMENT 24

PURPOSE

The purpose of this experiment is to familiarize you with the differential amplifier (diff-amp).

FORMAL OBJECTIVES

After completing this experiment, you will be able to:

- describe the DC operation of the diff-amp.
- describe the AC operation of the diff-amp.
- use the oscilloscope, function generator, and multimeter to test a diff-amp.

PREPARATIONS

In preparation for this experiment, you should study Sections 10-1 and 10-2 of *Exploring Electronic Devices.*

COMPONENTS and EQUIPMENT

1 - Bipolar DC source (±12 V, 20 mA)
1 - Plug-in experiment board (breadboard)
1 - Function generator with leads
3 - 10 kΩ, 1/4 W Resistors
1 - Multimeter with clip leads
1 - Dual-trace oscilloscope with leads
2 - 2N3904 NPN Transistors
2 - 1 kΩ, 1/4 W Resistors

PART ONE: DC CHARACTERISTICS OF THE DIFF-AMP

In this part of the experiment, you will assemble and test the differential amplifier of Figure 1. All test performed in this part of the experiment will be done with a multimeter and no signal shall be applied to the amplifier. Make sure you have reviewed Sections 10-1 and 10-2 of *Exploring Electronic Devices* before proceeding in this experiment.

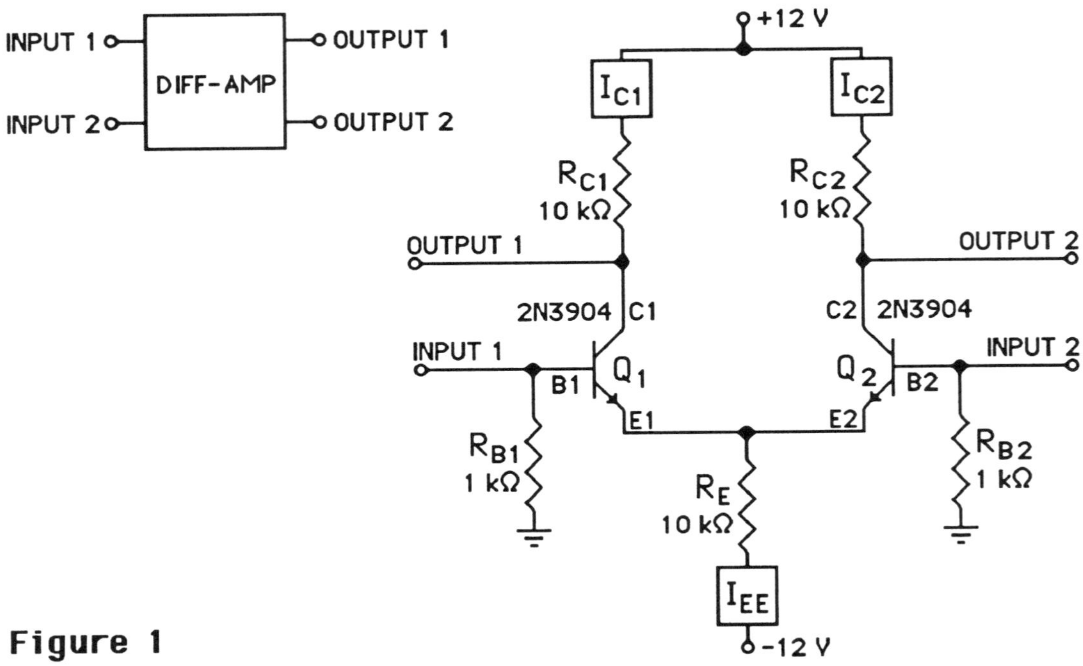

Figure 1

1. Assemble the diff-amp shown in Figure 1. Have someone check your work.
2. Calculate I_{EE}, I_{C1}, I_{C2}, V_{C1} and V_{C2}, V_{B1} and V_{B2}, V_E. All voltages are with respect to ground. Record your calculations on the answer sheet for Part One.
3. Apply ±12 V as indicated.
4. Use your multimeter as an ammeter and measure collector and emitter currents as indicated in the diagram. Record these readings on the answer sheet for Part One.
5. Use your multimeter as a voltmeter and measure V_{C1} and V_{C2}, V_{B1} and V_{B2}, and V_E. All voltages are with respect to ground. Record your readings on the answer sheet for Part One.
6. Remove power. Answer all remaining questions for Part One.

PART TWO: AC CHARACTERISTICS OF THE DIFF-AMP

In this part of the experiment, you will apply an AC signal to the inputs of the diff-amp and verify calculated values of voltage gain and common-mode rejection.

1. Calculate the approximate voltage amplification factor for each transistor where $A_V = R_C/2r'_e$. Record your answer on the answer sheet for Part Two.
2. Apply a 1 kHz, 100 $mV_{P\text{-}P}$ sine wave signal to Input 1 and use the oscilloscope to measure the output at Output 1. Record your reading on the answer sheet for Part Two and use it to determine the actual amplification factor.
3. Measure the peak-to-peak output voltage at Output 2 and record it on the answer sheet. Use this measured value to determine the amplification factor from Input 1 to Output 2.
4. Remove the signal from Input 1.
5. Apply a 1 kHz, 100 $mV_{P\text{-}P}$ sine wave signal to Input 2 and use the oscilloscope to measure the output at Output 2. Record your reading on the answer sheet for Part Two and use it to determine the actual amplification factor.
6. Measure the peak-to-peak output voltage at Output 1 and record it on the answer sheet. Use this measured value to determine the amplification factor from Input 2 to Output 1.
7. With the 100 $mV_{P\text{-}P}$ signal still applied to Input 2, use both channels of the oscilloscope to compare the signals at Output 1 and Output 2. Are these output signals in phase? (Place answer on answer sheet.)
8. Place a jumper wire between Input 1 and Input 2. This will cause the input signal to become common-mode.
9. Measure and record the resulting output signals.
10. Measure and record the signal that appears at the emitter of the transistors.
11. Remove power and answer all remaining questions for Part Two.

NAME:____________________CLASS:__________DATE:__________

PART ONE QUESTIONS

		Calculated Values	Measured Values
1.	I_{EE}	______	______
2.	I_{C1}	______	______
3.	I_{C2}	______	______
4.	V_{C1}	______	______
5.	V_{C2}	______	______
6.	V_{B1}	______	______
7.	V_{B2}	______	______
8.	V_E	______	______

9. What is the amount of emitter current through each transistor? ______________
10. Determine r'_e for the transistors. $r'_e = 25\ mV/I_{E(Q1\ or\ Q2)}$ = ______________Ω
11. $V_{CE(Q1)} \sim V_{CE(Q2)}$ = ______________

PART TWO QUESTIONS

1. $A_{V(Q1)}$ = ______________, $A_{V(Q2)}$ = ______________
2. Input 1 to Output 1: $\mathcal{V}_{out1}$ = ______________, A_V = ______________
3. Input 1 to Output 2: $\mathcal{V}_{out2}$ = ______________, A_V = ______________
4. Input 2 to Output 2: $\mathcal{V}_{out1}$ = ______________, A_V = ______________
5. Input 2 to Output 1: $\mathcal{V}_{out1}$ = ______________, A_V = ______________
6. Are the signals at Output 1 and 2 in phase? ______________
7. Input 1 and Input 2 tied together with signal applied:
   $\mathcal{V}_{out1}$ = ______________, A_V = ______________
   $\mathcal{V}_{out2}$ = ______________, A_V = ______________
8. Are the signals at Output 1 and 2 in phase with a common-mode signal applied? __________
9. $\mathcal{V}_E$ = ______________
10. Explain why the signal measured at the emitters of the transistors is approximately the same amplitude as the input signal when in common-mode.

____

____

____

____

Noninverting Op-Amp Circuits

EXPERIMENT 25

PURPOSE

The purpose of this experiment is to familiarize you with basic noninverting op-amp circuits and to demonstrate theory.

FORMAL OBJECTIVES

After completing this experiment, you will be able to:

- describe the operation of the unity-gain noninverting op-amp amplifier.
- determine the voltage amplification factor and decibel gain of any noninverting op-amp amplifier that uses voltage-divider negative feedback.

PREPARATIONS

In preparation for this experiment, you should study Sections 10-3 and 10-4 of *Exploring Electronic Devices.*

COMPONENTS and EQUIPMENT

1 - Dual-polarity DC source (±9 V to ±18 V)
1 - Plug-in experiment board (breadboard)
1 - Function generator with leads
1 - LF356 Op-amp
1 - 1 kΩ, 1/4 W Resistor
1 - Multimeter with clip leads
1 - Dual-trace oscilloscope with leads
1 - LM741 Op-amp
1 - 10 kΩ, 1/4 W Resistor

PART ONE: THE UNITY-GAIN NONINVERTING OP-AMP AMPLIFIER

In this part of the experiment, you will assemble and test the circuit shown in Figure 1. You should recognize this circuit as a unity-gain noninverting amplifier. Recall that this amplifier has no voltage gain or loss and the output signal is in phase with the input signal. Also, the input impedance to this amplifier is extremely high and its output impedance is extremely low because of the maximum negative feedback. Notice that the feedback voltage is equal to the output voltage and is nearly exactly equal to the input voltage. The slight difference that does exist between the input voltage and the feedback voltage is a very small difference voltage that is amplified by the op-amp to produce the output voltage where $v_{out} = v_{diff} \bullet A_{OL}$.

Figure 1

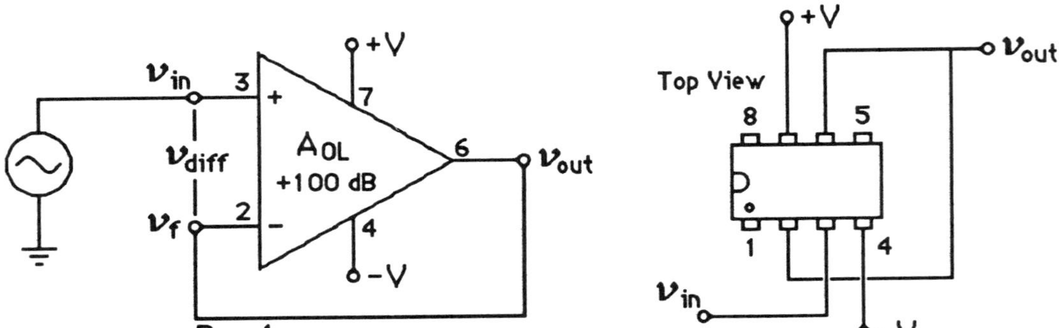

$B = 1$

$A_V = 1/B = 1/1 = 1 = B$

$v_{out} = v_{diff} \bullet A_{OL}$ and $v_{diff} = v_{out}/A_{OL}$

1. Assemble the circuit shown in Figure 1. All you will need is an op-amp and some jumper wire.
2. Apply ±9 V to ± 18 V to pins 7 and 4 of the op-amp. Do not apply an AC signal at this time.
3. Use your multimeter to measure the DC voltage at pins 3, 2, and 6. All measurements are with respect to a ground that is in common to the plus and minus DC supply. Record your measurements on the answer sheet for Part One.
4. Now connect the function generator to the input at pin 3 and set it to supply a 1 kHz, 1 $V_{P\text{-}P}$ sine wave.
5. Use the oscilloscope to measure the peak-to-peak output voltage at pin 6. Record this on the answer sheet and calculate the amplification factor and gain in decibels.
6. Use both channels of the oscilloscope to compare the phase of the input and output signals.
7. Increase the amplitude of the input signal until the output is at maximum voltage swing without distortion (no clipping). Record this maximum output voltage on the answer sheet.

8. Substitute the LF356 op-amp for the 741 op-amp in your circuit and determine whether there is any noticeable difference.

9. Remove power and answer all remaining questions for Part One.

PART TWO: THE NONINVERTING AMPLIFIER WITH GAIN

In this part of the experiment, you will construct the circuit of Figure 2 and demonstrate the effect of the feedback voltage divider on the amplification factor.

Figure 2

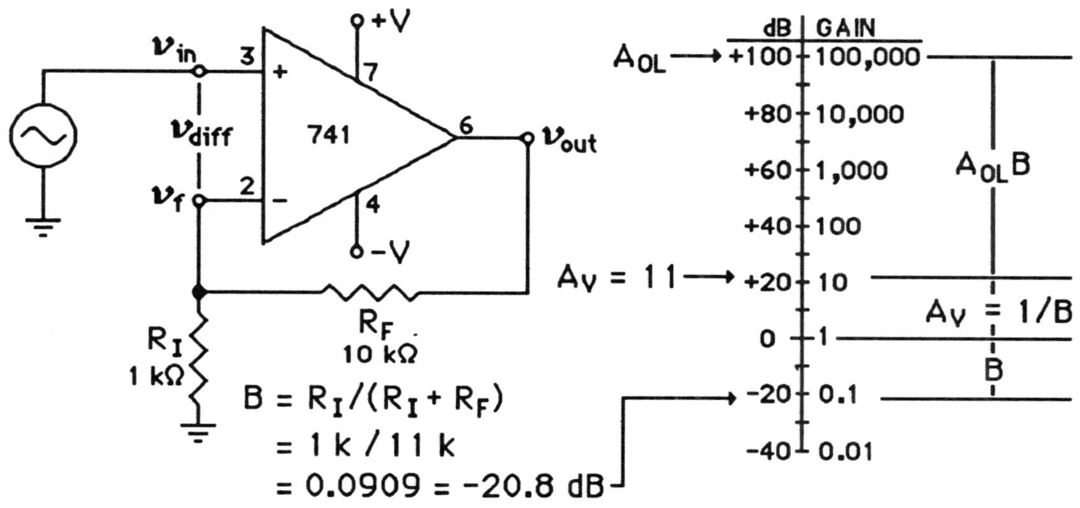

1. Assemble the circuit of Figure 2. Use the 741 op-amp.

2. Apply ±9 V to ± 18 V to pins 7 and 4 of the op-amp. Do not apply an AC signal at this time.

3. Use your multimeter to measure the DC voltage at pins 3, 2, and 6. All measurements are with respect to a ground that is in common to the plus and minus DC supply. Record your measurements on the answer sheet for Part Two.

4. Now connect the function generator to the input at pin 3 and set it to supply a 1 kHz, 1 V_{P-P} sine wave.

5. Use the oscilloscope to measure the peak-to-peak output voltage at pin 6. Record this on the answer sheet and calculate the amplification factor and gain in decibels.

6. Use both channels of the oscilloscope to compare the phase of the input and output signals.

7. Increase the amplitude of the input signal until the output is at maximum voltage swing without distortion (no clipping). Record this maximum output voltage on the answer sheet.

8. Substitute the LF356 op-amp for the 741 op-amp in your circuit and determine whether there is any noticeable difference.

9. Substitute a different value resistor for R_F in your circuit and determine the voltage feedback ratio B, A_V, and gain in dB. Verify your calculations experimentally.

10. Remove power and answer all remaining questions for Part Two.

NOTES AND CALCULATIONS

NAME:________________________________**CLASS:**__________**DATE:**__________

PART ONE QUESTIONS

1. $V_{pin\ 2}$ = ____________ V, $V_{pin\ 3}$ = ____________ V, $V_{pin\ 6}$ = ____________ V

2. $\mathcal{V}_{out}$ = ____________ $V_{P\text{-}P}$, $A_V = \mathcal{V}_{out}/\mathcal{V}_{in}$ = ____________________

   Gain in dB = ________________________ = __________ dB

3. Phase relationship? ____________________________

4. Maximum output voltage swing without clipping = ____________ $V_{P\text{-}P}$

5. If the output voltage is set at 1 $V_{P\text{-}P}$ and the open-loop gain of the op-amp is 100 dB, what is the value of the difference voltage between pins 2 and 3? Show all work.

   ____

   ____

6. Was there any noticeable difference when you substituted the LF356 for the 741? ________
   If so, what? ________________________________

PART TWO QUESTIONS

1. $V_{pin\ 2}$ = ____________ V, $V_{pin\ 3}$ = ____________ V, $V_{pin\ 6}$ = ____________ V

2. $\mathcal{V}_{out}$ = ____________ $V_{P\text{-}P}$, $A_V = \mathcal{V}_{out}/\mathcal{V}_{in}$ = ____________________

   Gain in dB = ________________________ = __________ dB

3. Phase relationship? ____________________________

4. Maximum output voltage swing without clipping = ____________ $V_{P\text{-}P}$

5. If the output voltage is set at 1 $V_{P\text{-}P}$ and the open-loop gain of the op-amp is 100 dB, what is the value of the difference voltage between pins 2 and 3? Show all work.

   ____

   ____

6. If the output voltage is set at 1 $V_{P\text{-}P}$, what is the value of the feedback voltage at pins 2? Show all work.

   ____

7. Was there any noticeable difference when you substituted the LF356 for the 741? ________

   If so, what? ____

8. What value resistor did you substitute for R_F? ____________ Ω

9. Using your resistor: B = ____

   A_V = ____

   #dB gain = ____

NOTES AND CALCULATIONS

Inverting Op-Amp Circuits

EXPERIMENT 26

PURPOSE

The purpose of this experiment is to familiarize you with the basic inverting op-amp amplifier. You will use this experiment to verify theory regarding voltage amplification factor and virtual ground.

FORMAL OBJECTIVES

After completing this experiment, you will be able to:

- determine the amplification factor for an inverting operational amplifier when the feedback and input resistances are known.
- explain the meaning of virtual ground at the inverting input.
- use a multimeter, function generator, and oscilloscope to analyze an inverting op-amp circuit.

PREPARATIONS

In preparation for this experiment, you should study Section 10-4 of *Exploring Electronic Devices*.

COMPONENTS and EQUIPMENT

1 - Dual-polarity DC source (±9 V to ±18 V)
1 - Plug-in experiment board (breadboard)
1 - Function generator with leads
1 - LF356 Op-amp
1 - 1 kΩ, 1/4 W Resistor
1 - Multimeter with clip leads
1 - Dual-trace oscilloscope with leads
1 - LM741 Op-amp
1 - 10 kΩ, 1/4 W Resistor

DISCUSSION

In the following steps, you will perform a DC and AC analysis of the inverting amplifier shown in Figure 1. In so doing, you will verify virtual ground at pin 2 and amplification factor. Recall that a virtual ground is a point in a circuit at which the voltage is always equal to, or nearly equal to, the ground potential (near 0 V).

Figure 1

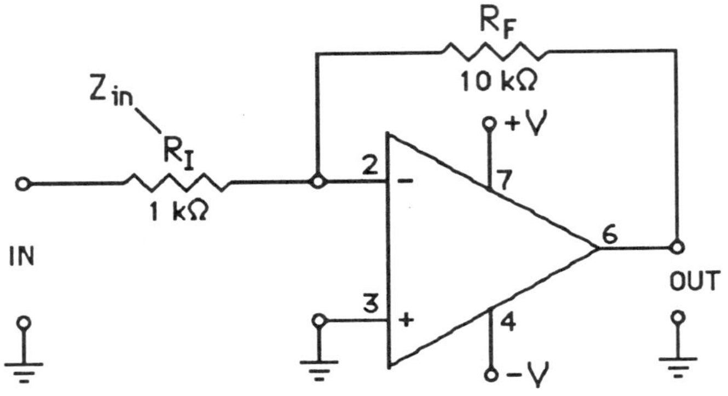

PROCEDURE

1. Construct the circuit shown in Figure 1. Use the 741 op-amp.
2. Before applying power, predict the DC voltage at pin 2 and 6. Also, calculate the amplification factor A_V where $A_V = R_F/R_I$. Record your answers on the answer sheet.
3. Apply bipolar voltage to your circuit in the range of ±9 V to ±18 V (pins 7 and 4).
4. Measure and record the DC voltages at pins 2 and 6.
5. Apply +0.5 VDC to the input. Measure and record the DC voltages at pins 2 and 6 on the answer sheet. Also, determine the voltage amplification factor ($V_{out}/V_{in} = V_{out}/0.5$ V).
6. Disconnect the +0.5 V from the input.
7. Use the function generator to apply a 1 kHz, 1 $V_{P\text{-}P}$ sine wave to the input of your circuit.

   CAUTION! Make sure the sine wave is zero-referenced - no DC offset.
8. With the 1 $V_{P\text{-}P}$ sine wave applied, use the oscilloscope to measure the AC voltage at pin 2 and at pin 6. Record your readings on the answer sheet.
9. Use both channels of the oscilloscope to determine the phase relationship of the input and output signals.
10. Substitute the LF356 for the 741 op-amp and determine if there is any change in A_V.
11. Remove power and answer all remaining questions.

NAME:____________________ CLASS:__________ DATE:__________

QUESTIONS

1. Predicted values: V_{pin2} = 0v V, V_{pin6} = -10v V, A_V = 10

2. Measured values: V_{pin2} = 0v V, V_{pin6} = -9.8v V

3. With +0.5 V DC in: V_{pin2} = 0v V, V_{pin6} = -5v V, A_V = 20

4. With +0.5 VDC applied to the input is the voltage at the output (a) positive or (b) negative? ___

5. Would you say the input and output voltages are (a) in phase or (b) 180° out-of-phase? ___

6. With 1 $V_{P\text{-}P}$ in: V_{pin2} = 0v V, V_{pin6} = ______ V, A_V = ______

7. Would you say the input and output voltages are (a) in phase or (b) 180° out-of-phase? ___

8. Was there any change in A_V when you substituted the LF356 for the 741? ______
   Why? ____________________

9. How does the data collected in this experiment support the idea that pin 2 of the op-amp is a virtual ground?

   ____

   ____

10. When +0.5 VDC was applied to the input, what was the amount of current through R_I and through R_F? Show your calculations.
    I_{RI} = ____________________
    I_{RF} = ____________________

11. When 1 V_{P-P} was applied to the input, what was the amount of AC voltage across R_I and across R_F? Show your calculations.

v_{RI} = $I_{R_I} \times R_I$ = 1mApp × 1000 = 1V

v_{RF} = ______

12. Use the AC voltages calculated in #11 to determine the AC current flowing through R_I and through R_F. Show your calculations.

i_{RI} = ______

i_{RF} = ______

13. What value of R_F would you need if you wanted the A_V of the amplifier to be 33? ______ (Assume $R_I = 1\ k\Omega$.)

14. What value of R_I would you need if you wanted the A_V of the amplifier to be 1? ______ (Assume $R_F = 10\ k\Omega$.)

NOTES AND CALCULATIONS

Op-Amp Slew Rate and Frequency Response

EXPERIMENT 27

PURPOSE

The purpose of this experiment is to familiarize you with op-amp slew rate, power bandwidth, and small-signal bandwidth.

FORMAL OBJECTIVES

After completing this experiment, you will be able to:

- use the oscilloscope and function generator to determine op-amp slew rate (SR).
- describe how the slew rate of an op-amp effects its large-signal bandwidth.
- describe how the small-signal bandwidth of an op-amp is effected by A_V.

PREPARATIONS

In preparation for this experiment, you should study Sections 10-3, 11-1, and 11-2 of *Exploring Electronic Devices*.

COMPONENTS and EQUIPMENT

1 - Dual-polarity DC source (±12 V)
1 - Dual-trace oscilloscope with leads
1 - LM741 Op-amp
1 - 47 kΩ, 1/4 W Resistor
1 - 4.7 kΩ, 1/4 W Resistor
1 - Plug-in experiment board (breadboard)
1 - Function generator with leads
1 - LF356 Op-amp
1 - 10 kΩ, 1/4 W Resistor
2 - 1 kΩ, 1/4 W Resistors

PART ONE: OP-AMP SLEW RATE

In this part of the experiment, you will determine the slew rate of the 741 and LF356 op-amps. The noninverting unity-gain configuration of Figure 1 will be used.

Figure 1

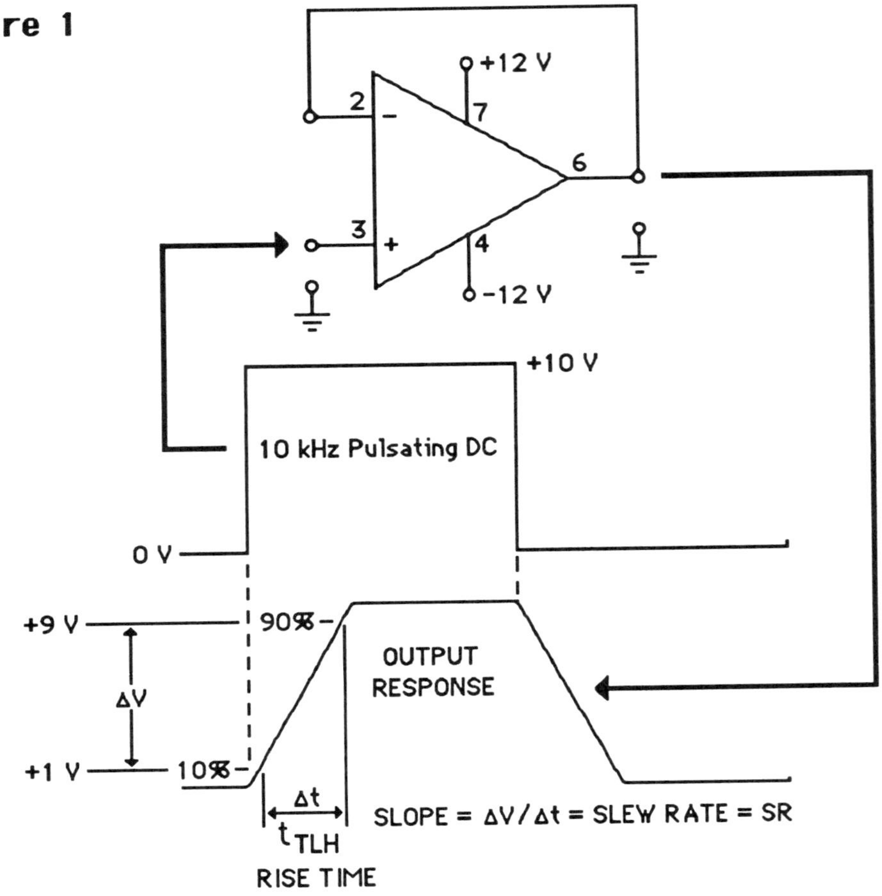

1. Wire the 741 op-amp as shown in Figure 1. Apply ±12 V as shown.

2. Set your function generator for a +10 V, 10 kHz pulsating DC square wave as shown. You may have to adjust the DC offset control on the function generator along with the amplitude control to obtain the desired signal. Connect this signal to the input of your op-amp.

3. Use channel 1 of your oscilloscope to observe the input pulses and channel 2 to observe the output signal. Set the Time/Div control to 10 or 20 μs/div.

4. Determine the slew rate of the op-amp from the rising edge of the output waveform where Δt is the time it takes for the output to rise from +1 to +9 V (8 V). SR = 8 V/Δt. Record all answers on the answer sheet for Part One.

5. Replace the 741 with the LF356 op-amp and determine its slew rate. You may have to increase the frequency of the input square wave and reduce the Time/Div setting of the oscilloscope.

6. Remove power and answer all remaining questions for Part One.

PART TWO: SLEW RATE AND POWER BANDWIDTH

In this part of the experiment, you will observe the effects slew rate has on the power bandwidth of the op-amp. Recall that large-signal slew-rate distortion is caused by internal capacitance of the op-amp that is unable to charge and discharge as quickly as the input signal rises and falls.

Figure 2

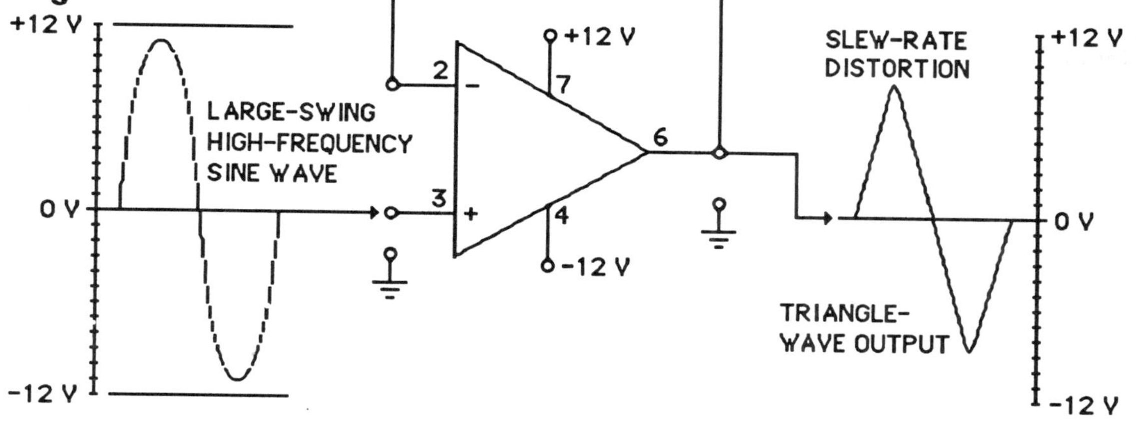

1. Wire the 741 op-amp as shown in Figure 2. Apply ±12 V as shown.

2. Set your function generator for a 22 V_{P-P} , 20 kHz sine wave as shown. Make sure any DC offset is turned off. Connect this signal to the input of your op-amp.

3. Use channel 1 of your oscilloscope to observe the input sine wave and channel 2 to observe the output signal. Set the Time/Div control to 10 or 20 μs/div.

4. You should now see that the output waveform is not a sine wave and its amplitude is less than 22 V_{P-P}. This is called slew-rate distortion.

5. Slowly decrease the *frequency* of the input signal until the output signal appears to be the same as the input signal. Measure and record this frequency on the answer sheet for Part Two.

6. Set your function generator for a 10 V_{P-P} , 200 kHz sine wave.

7. You should now see that the output waveform is not a sine wave and its amplitude is less than 10 V_{P-P}.

8. Slowly decrease the *frequency* of the input signal until the output signal appears to be the same as the input signal. Measure and record this frequency on the answer sheet for Part Two.

9. Replace the 741 with the LF356 op-amp and repeat Steps 2 through 8. Do you notice any performance difference as compared to the 741?

10. Remove power and answer all remaining questions for Part Two.

PART THREE: A_V AND SMALL-SIGNAL BANDWIDTH

In this part of the experiment, you will explore the relationship between amplification factor and bandwidth for an op-amp circuit. The inverting configuration of Figure 3 will be used. The method used here is to set the amplification factor to different values by changing R_F and observing the change in the upper cutoff frequency (f_{co}). Recall that $f_{co} = f_T/A_V$ and $f_T = A_V \bullet f_{co}$ where f_T is the unity-gain frequency, the frequency at which the open-loop gain is one (1). You should observe that the upper cutoff frequency is greatest when A_V is set to one (1).

Figure 3

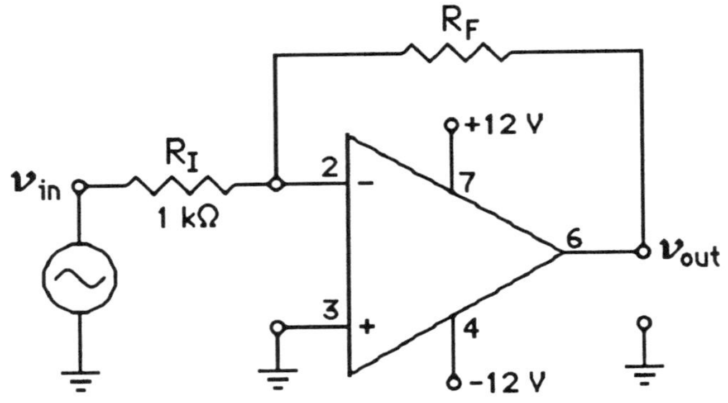

$$A_V = v_{out}/v_{in} = R_F/R_I$$

1. Configure your op-amp circuit as shown in Figure 3 using the 741. Let $R_I = R_F = 1\ k\Omega$.

2. Set the function generator for a 1 kHz sine wave. Adjust the function generator's amplitude control so the signal at the output of the op-amp (pin 6) is 1 $V_{P\text{-}P}$.

3. Set the Volts/Div control for the oscilloscope channel you are using to 0.2 V/div. The 1 $V_{P\text{-}P}$ output signal should occupy five vertical divisions.

4. Increase the frequency of the input signal until the output signal's amplitude drops to 0.707 $V_{P\text{-}P}$ (3.5 vertical divisions). This is the upper cutoff frequency at which the output voltage is equal to $0.707 \bullet V_{(max)}$, where $V_{(max)}$ is 1 $V_{P\text{-}P}$ in this case.

5. Adjust the oscilloscope's Time/Div control so the time (T) for one cycle can be measured. Calculate the cutoff frequency where f = 1/T. Record your answer on the answer sheet for Part Three.

6. Repeat Steps 2 through 5 three more times substituting the 4.7 kΩ, 10 kΩ, and 47 kΩ resistor for RF each time.

7. Remove all power and answer any remaining questions on the answer sheet.

NAME:______________________________ CLASS:__________ DATE:__________

PART ONE QUESTIONS

1. Using the 741 op-amp: $SR = \Delta V/\Delta t = 8\ V$ / ______ = __________ V/μs
2. Using the LF356 op-amp: $SR = \Delta V/\Delta t = 8\ V$ / ______ = __________ V/μs
3. Which op-amp has the higher slew rate? ______________

PART TWO QUESTIONS

1. Using the 741 op-amp: What was the highest undistorted large-signal frequency as determined in Step 5? ____________
2. Using the 741 op-amp: What was the highest undistorted large-signal frequency as determined in Step 8? ____________
3. How did decreasing the input signal effect the power bandwidth of the 741 circuit?

   ____
4. Using the LF356 op-amp: What was the highest undistorted large-signal frequency as determined in Step 5? ____________
5. Using the LF356 op-amp: What was the highest undistorted large-signal frequency as determined in Step 8? ____________
6. Which of the two op-amps has the widest power bandwidth? ___________
7. From your reading in Chapter 11 of Exploring Electronic Devices, what are two precautions that can be taken to avoid slew-rate distortion?

   1. ____

   2. ____

PART THREE QUESTIONS

1. $R_I = 1\ k\Omega$, $R_F = 1\ k\Omega$, A_V = ___________, f_{co} = ___________

   Gain-Bandwidth Product = $f_T = A_V \cdot f_{co}$ = ______________________________
2. $R_I = 1\ k\Omega$, $R_F = 4.7\ k\Omega$, A_V = ___________, f_{co} = ___________

   Gain-Bandwidth Product = $f_T = A_V \cdot f_{co}$ = ______________________________

3. $R_I = 1\ k\Omega$, $R_F = 10\ k\Omega$, $A_V =$ __________, $f_{co} =$ __________

   Gain-Bandwidth Product $= f_T = A_V \bullet f_{co} =$ ____________________

4. $R_I = 1\ k\Omega$, $R_F = 47\ k\Omega$, $A_V =$ __________, $f_{co} =$ __________

   Gain-Bandwidth Product $= f_T = A_V \bullet f_{co} =$ ____________________

5. From the data collected above, describe the relationship between amplification factor (A_V) and upper cutoff frequency (f_{co}).

   ____

6. How can the upper cutoff frequency be determined mathematically if A_V and f_T are known?

   ____

NOTES AND CALCULATIONS

Single-Supply Op-Amp Circuits

EXPERIMENT 28

PURPOSE

The purpose of this experiment is to demonstrate how to operate an op-amp from a single DC supply. Both the noninverting and inverting configurations will be explored.

FORMAL OBJECTIVES

After completing this experiment, you will be able to:

- build single-supply noninverting amplifiers.
- build single-supply inverting amplifiers.

PREPARATIONS

In preparation for this experiment, you should study Section 12-1 of *Exploring Electronic Devices.*

COMPONENTS and EQUIPMENT

1 - DC source (12 V, 20 mA capability)
1 - Plug-in experiment board (breadboard)
1 - Function generator with leads
1 - LM741 Op-Amp
1 - 1 μF, 35 V Electrolytic capacitor
1 - 10 kΩ, 1/4 W Resistor
1 - Digital multimeter with clip leads
1 - Dual-trace oscilloscope with leads
1 - LF356 Op-Amp
2 - 10 μF, 35 V Electrolytic capacitors
3 - 100 kΩ, 1/4 W Resistors
1 - 2 kΩ, 1/4 W Resistor

PART ONE: THE SINGLE-SUPPLY NONINVERTING AMPLIFIER

In this part of the experiment, you will assemble and test the single-supply noninverting amplifier shown in Figure 1. Recall from your reading that $Z_{in} = R_D/2$ and $A_V = (R_F + R_I)/R_I$.

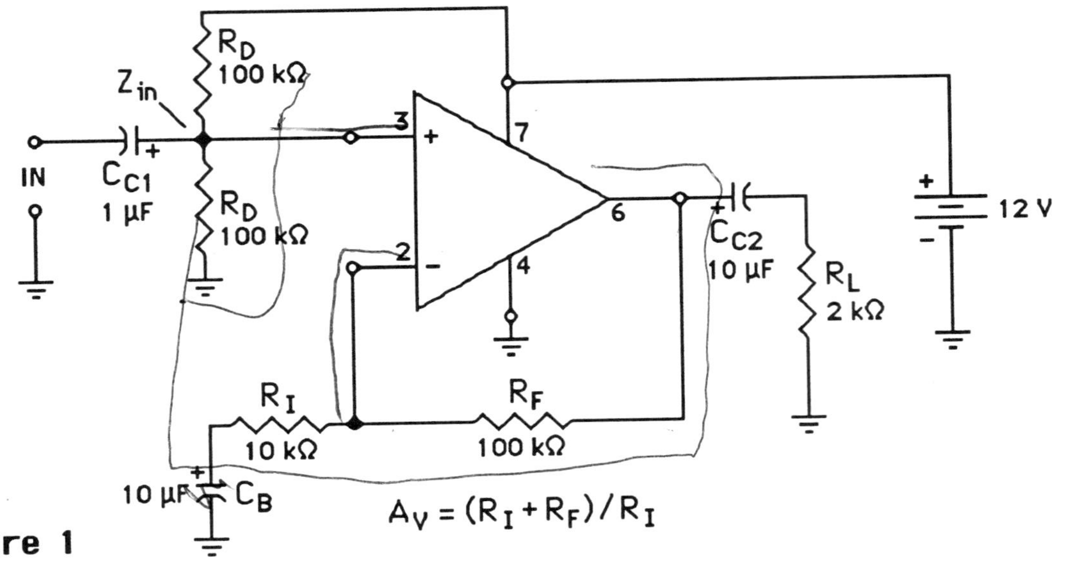

Figure 1

1. Assemble the circuit shown in Figure 1 using the 741 op-amp.
2. Apply 12 VDC as shown.
3. Use your multimeter to measure the DC voltages at pins 2, 3, and 6. Record your measurements on the answer sheet for Part One.
4. Use your multimeter to measure the DC voltage across C_B and R_L. Record your measurements on the answer sheet for Part One.
5. Apply a 1 kHz, 500 $mV_{P\text{-}P}$ sine wave to the input.
6. Use both channels of the oscilloscope to observe and compare the input and output signals. Determine the amplification factor of this circuit using $(R_F + R_I)/R_I$ and V_{out}/V_{in}. Record your calculations on the answer sheet for Part One.
7. Increase the amplitude of the input signal and determine the maximum undistorted (unclipped) output signal. Record your measurement on the answer sheet for Part One.
8. Repeat Steps 2 through 7 using the LF356 op-amp.
9. Remove power. Answer any remaining questions for Part One.

PART TWO: THE SINGLE-SUPPLY INVERTING AMPLIFIER

In this part of the experiment, you will assemble and test the single-supply inverting amplifier shown in Figure 2. Recall from your reading that $Z_{in} = R_I$ and $A_V = R_F/R_I$.

Figure 2

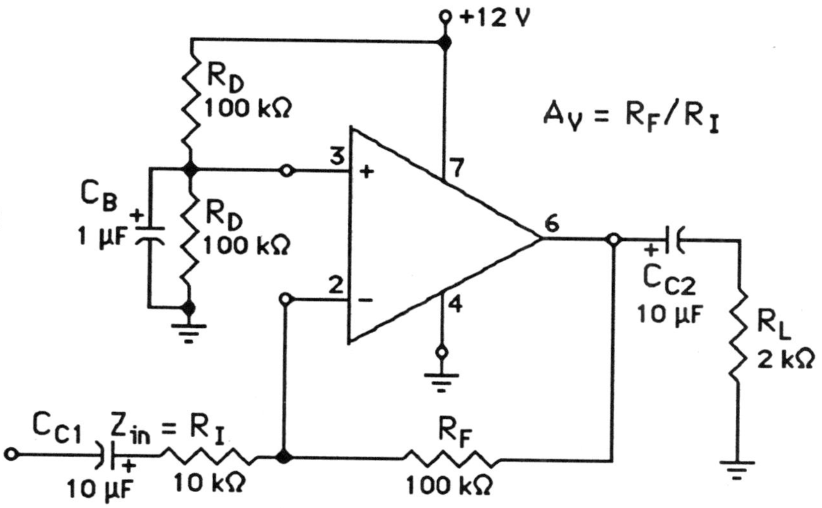

1. Assemble the circuit shown in Figure 2 using the 741 op-amp.
2. Apply 12 VDC as shown.
3. Use your multimeter to measure the DC voltages at pins 2, 3, and 6. Record your measurements on the answer sheet for Part Two.
4. Use your multimeter to measure the DC voltage across C_B and R_L. Record your measurements on the answer sheet for Part Two.
5. Apply a 1 kHz, 500 $mV_{P\text{-}P}$ sine wave to the input.
6. Use both channels of the oscilloscope to observe and compare the input and output signals. Determine the amplification factor of this circuit using R_F/R_I and V_{out}/V_{in}. Record your calculations on the answer sheet for Part Two.
7. Increase the amplitude of the input signal and determine the maximum undistorted (unclipped) output signal. Record your measurement on the answer sheet for Part Two.
8. Repeat Steps 2 through 7 using the LF356 op-amp.
9. Remove power. Answer any remaining questions for Part Two.

NAME:______________________________ CLASS:__________ DATE:__________

PART ONE QUESTIONS

Using the 741 op-amp:

1. V_{pin2} = __________V, V_{pin3} = __________V, V_{pin6} = __________V
2. V_{CB} = __________V, V_{RL} = __________V
3. Using R_F and R_I: A_V = ____________________________ = __________

   Using $\mathcal{V}_{out}$ and $\mathcal{V}_{in}$: A_V = ____________________________ = __________
4. Maximum output voltage = __________$V_{P\text{-}P}$

Using the LF356 op-amp:

5. V_{pin2} = __________V, V_{pin3} = __________V, V_{pin6} = __________V
6. V_{CB} = __________V, V_{RL} = __________V
7. Using R_F and R_I: A_V = ____________________________ = __________

   Using $\mathcal{V}_{out}$ and $\mathcal{V}_{in}$: A_V = ____________________________ = __________
8. Maximum output voltage = __________$V_{P\text{-}P}$

9. Was there any difference in performance between the 741 and the LF356 op-amp? ________
10. In a single-supply noninverting op-amp circuit, what determines the DC voltage at pin 6?

    ____
11. What determines the input impedance for the single-supply noninverting op-amp circuit?

    ____

12. Draw and label a single-supply noninverting op-amp circuit that has an input impedance equal to 90 kΩ and a voltage amplification factor equal to 23. The maximum unclipped output voltage shall be approximately 12 to 13 $V_{P\text{-}P}$.

PART TWO QUESTIONS

Using the 741 op-amp:

1. V_{pin2} = __________V, V_{pin3} = __________V, V_{pin6} = __________V
2. V_{CB} = __________V, V_{RL} = __________V
3. Using R_F and R_I: A_V = ____________________ = __________

   Using $\mathcal{V}_{out}$ and $\mathcal{V}_{in}$: A_V = ____________________ = __________
4. Maximum output voltage = __________$V_{P\text{-}P}$

Using the LF356 op-amp:

5. V_{pin2} = __________V, V_{pin3} = __________V, V_{pin6} = __________V
6. V_{CB} = __________V, V_{RL} = __________V
7. Using R_F and R_I: A_V = ____________________ = __________

   Using $\mathcal{V}_{out}$ and $\mathcal{V}_{in}$: A_V = ____________________ = __________
8. Maximum output voltage = __________$V_{P\text{-}P}$
9. Was there any difference in performance between the 741 and the LF356 op-amp? __________

10. In a single-supply inverting op-amp circuit, what determines the DC voltage at pin 6?

____

11. What determines the input impedance for the single-supply inverting op-amp circuit?

____

12. Draw and label a single-supply inverting op-amp circuit that has an input impedance equal to 22 kΩ and a voltage amplification factor equal to 10. The maximum unclipped output voltage shall be approximately 10 $V_{P\text{-}P}$.

NOTES AND CALCULATIONS

Power-Boosting an Op-Amp

EXPERIMENT **29**

PURPOSE

The purpose of this experiment is to demonstrate the manner in which the output of an op-amp can be power-boosted using a complementary transistor pair.

FORMAL OBJECTIVES

After completing this experiment, you will be able to:

• construct and test a power-boosted op-amp circuit with a complementary-pair push-pull transistor output.

PREPARATIONS

In preparation for this experiment, you should study Section 12-4 of *Exploring Electronic Devices*.

COMPONENTS and EQUIPMENT

1 - Dual-polarity DC source (±12 V)
1 - Plug-in experiment board (breadboard)
1 - Function generator with leads
1 - 2N3904 NPN Transistor
1 - 10 kΩ, 1/4 W Resistor
1 - 51 Ω, 1 W Resistor
1 - Multimeter with clip leads
1 - Dual-trace oscilloscope with leads
1 - LM741 Op-amp
1 - 2N3906 PNP Transistor
1 - 100 Ω, 1/4 W Resistor
2 - 100 µF, 35 V Electrolytic capacitors

DISCUSSION

In the following steps, you will assemble a power-boosted op-amp circuit and investigate its performance. You will begin with a deliberately inadequate circuit (Figure 1) to demonstrate the problem of crossover distortion. Then, you will make the necessary modifications to the circuit to achieve acceptable performance.

Figure 1

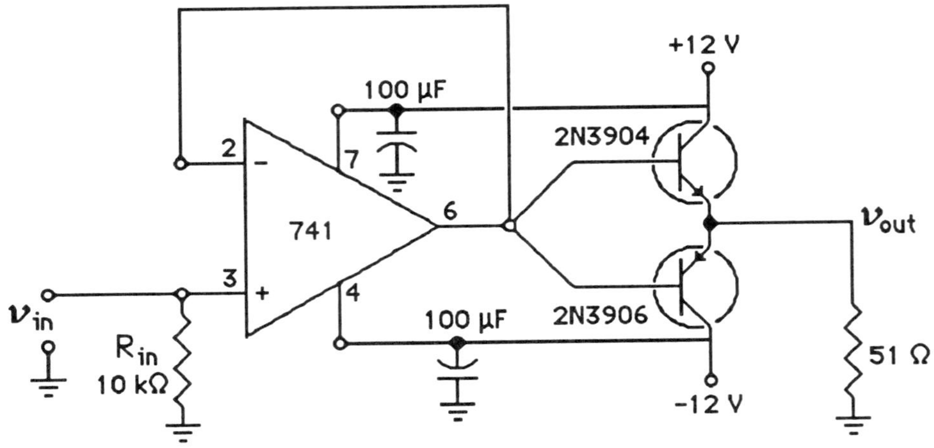

PROCEDURE

1. Construct the circuit shown in Figure 1. You may use any pin-compatible op-amp such as the 741 or the LF356. Have someone check your wiring.

2. Apply ±12 V as shown and use your multimeter to measure the DC voltage at pins 2, 3, and 6. Also, measure the DC voltage at the emitter of the transistors (across the 51 Ω load resistor). All voltage measurements are made with respect to the common power supply ground. Record your readings on the answer sheet.

3. Connect the function generator to the input of your circuit. Set the function generator to produce a 1 kHz sine wave that is 1 $V_{P\text{-}P}$ (no DC offset).

4. Use both channels of the oscilloscope to observe and compare the input and output signals. Describe what you see on the answer sheet.

5. Increase the amplitude of the input signal to 5 $V_{P\text{-}P}$. Use both channels of the oscilloscope to observe and compare the input and output signals. Describe what you see on the answer sheet.

6. Remove power.

7. Modify your circuit as shown in Figure 2. Add the 100 Ω resistor and move the negative-feedback jumper from pin 6 to the emitters of the transistors.

8. Repeat Steps 2 through 4.

Figure 2

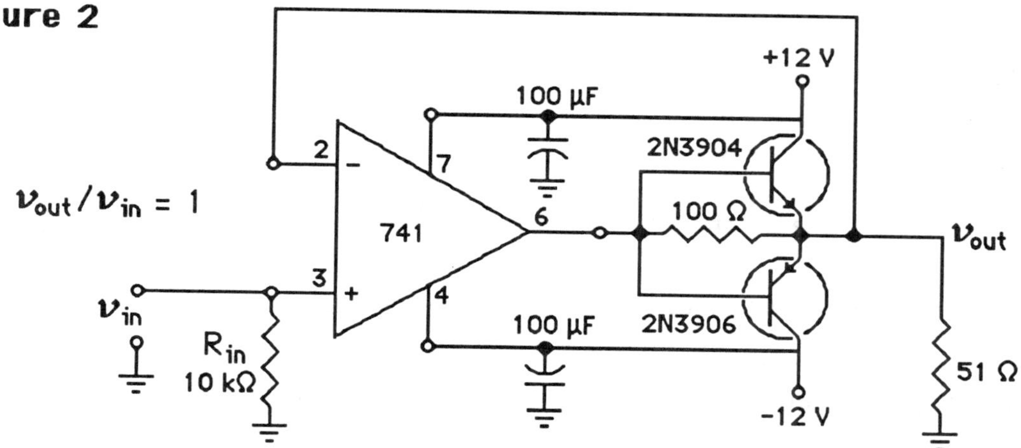

9. Determine the voltage amplification factor for your circuit.
10. *Briefly* increase the input signal to determine the maximum unclipped output signal.
11. Remove power and answer all remaining questions.

NAME:______________________________ CLASS:__________ DATE:__________

QUESTIONS

Circuit of Figure 1

1. V_{pin2} = __________V, V_{pin3} = __________V, V_{pin6} = __________V
   V_{load} = __________V
2. Are the output transistors conducting? ________ Why? ______________________
   ____
3. With a 1 V_{P-P} signal applied, what do you see in the output and why?
   ____
   ____
4. With a 5 V_{P-P} signal applied, what do you see in the output and why?
   ____
   ____

Circuit Modified - Figure 2

5. V_{pin2} = __________V, V_{pin3} = __________V, V_{pin6} = __________V
   V_{load} = __________V
6. Are the output transistors conducting? ________ Why? ______________________
   ____
7. With a 1 V_{P-P} signal applied, what do you see in the output and why?
   ____
   ____
8. A_V = ______________
9. Maximum output voltage swing = ____________V_{P-P}.
10. Explain why the crossover distortion disappeared when the circuit was modified to match Figure 2.
    ____
    ____
    ____

Precision Rectifiers

EXPERIMENT
30

PURPOSE

The purpose of this experiment is to familiarize you with precision rectifiers using op-amps. Here, you will investigate precision half-wave and full-wave rectifiers.

FORMAL OBJECTIVES

After completing this experiment, you will be able to:

- describe the advantage of using a precision rectifier as compared to a simple diode rectifier.
- wire an op-amp as a precision half-wave or full-wave rectifier.
- use the oscilloscope and function generator to test a precision rectifier.

PREPARATIONS

In preparation for this experiment, you should study Section 12-7 of *Exploring Electronic Devices*.

COMPONENTS and EQUIPMENT

1 - Bipolar DC source (±12 V)
1 - Plug-in experiment board
1 - Function generator with leads
2 - 1N914 Diodes
1 - 10 kΩ, 1/4 W Resistor
1 - Multimeter with clip leads
1 - Dual-trace oscilloscope with leads
2 - LF356 Op-amps
4 - 20 kΩ, 1/4 W Resistors

PART ONE: THE PRECISION HALF-WAVE RECTIFIER

In this part of the experiment, you will assemble and test the precision half-wave rectifier of Figure 1. With an AC signal applied to the input, a half-wave pulsating DC will appear in the output. Notice from the polarity of the diode that the output will be a positive pulsating DC. Also note that maximum negative feedback is supplied to the inverting input through the forward-biased diode during the positive alternations of the output voltage at pin 6. That makes this a unity-gain amplifier for all positive alternations.

1. Assemble the precision half-wave rectifier of Figure 1. Have someone check your wiring.
2. Set your function generator to produce a 1 kHz sine wave of 200 mV_{P-P}. Make sure there is no DC offset.
3. Apply ±12 V as shown. Connect the test signal to the input of the op-amp.
4. Use both channels of your oscilloscope to observe and compare the input and output signals.

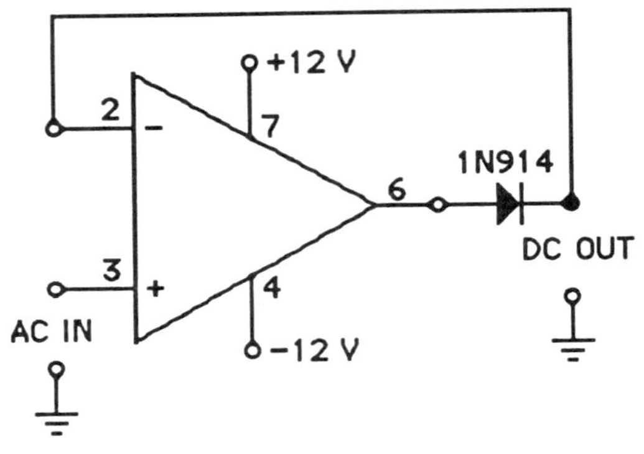

Figure 1

5. Vary the amplitude of the input signal from approximately 100 mV_{P-P} to 20 V_{P-P} while comparing the input and output signals.
6. Remove power and answer all questions for Part One.

PART TWO: THE PRECISION FULL-WAVE RECTIFIER

In this part of the experiment, you will assemble and test the precision full-wave rectifier of Figure 2. As you consider Figure 2, notice that the first stage is a precision half-wave rectifier that has unity gain for all negative alternations which cause the output diode D_1 to conduct where $A_V = R_2/R_1 = 20\ k\Omega/20\ k\Omega = 1$. For all positive alternations of the output at pin 6 of the rectifier stage, the feedback diode D_2 conducts ensuring a very low A_V. This prevents the output from saturating on every positive alternation at pin 6 of U_1.

The second stage of this circuit is a dual-input summing amplifier. One input, via R_5, has an A_V of 1 equal to R_4/R_5. The second input, via R_3, has an A_V of 2 equal to R_4/R_3. The negative alternations of an input AC signal at TP1 is passed through R_5 to the adder circuit which inverts it with an amplification factor of 1 to the output at TP3. Thus, a negative alternation of -1 V_P at TP1 will appear as a positive alternation of +1 V_P at TP3. Conversely, all positive alternations at TP1 will appear at TP2 as negative alternations to then be amplified by a factor of 2 at the summing amplifier. Thus, a positive alternation of 1 V_P at TP1 will appear as a negative alternation of 1 V_P at TP2 and finally appear as a positive alternation of 2 V_P at TP3. However, the same 1 V_P positive alternation at TP1 is also being amplified by a factor of 1 via R_5 and is inverted by the summing amplifier. Therefore, the actual output at TP3 is the difference between the two would-be output signals where $V_{TP3} = 2\ V_P - 1\ V_P = 1\ V_P$. The result of all this is precision full-wave rectification where each alternation of the AC input signal is converted to positive pulsating DC.

Figure 2

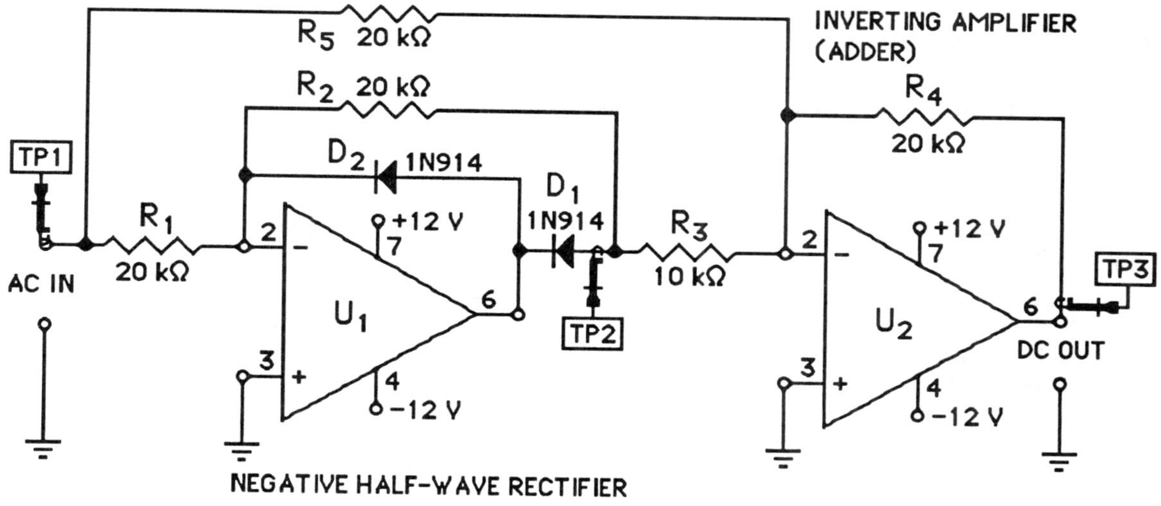

1. Assemble the precision full-wave rectifier of Figure 2. Definitely have someone check your work.
2. Set your function generator to produce a 1 kHz sine wave of 200 mV_{P-P}. Make sure there is no DC offset.
3. Apply ±12 V as shown. Connect the test signal to the input of the op-amp.
4. Use both channels of your oscilloscope to observe and compare the input TP1 and output TP3 signals.
5. Vary the amplitude of the input signal from approximately 100 mV_{P-P} to 20 V_{P-P} while comparing the input and output signals.
6. Now, use your oscilloscope to compare the signals at TP1 and TP2. Look for any difference in peak voltage and decide which alternation of the input signal is responsible for the signal at TP2.
7. Next, use your oscilloscope to compare the signals at TP2 and TP3.
8. Momentarily remove R_5 and observe what happens to the output at TP3.
9. Replace R_5. Momentarily remove R3 and observe what happens to the output at TP3.
10. Remove power and answer all questions for Part Two.

NAME:____________________ CLASS:__________ DATE:__________

PART ONE QUESTIONS

1. Is the precision rectifier of Figure 1 a precision (a) half-wave, or (b) full-wave rectifier? _____

2. Is the circuit of Figure 1 a precision (a) positive rectifier, or (b) negative rectifier? _____

3. How can the circuit of Figure 1 be modified to produce a pulsating DC output of the opposite polarity?

4. How is the performance of this precision rectifier different than that of a single-diode half-wave rectifier?

5. How does the peak amplitude of the output pulsating DC compare to the peak amplitude of the input AC signal? _____

6. Explain why this precision rectifier can rectify AC signals less than 0.7 V_P.

PART TWO QUESTIONS

1. The circuit of Figure 2 is a
   a. precision positive half-wave rectifier.
   b. precision positive full-wave rectifier.
   c. precision negative half-wave rectifier.
   d. precision negative full-wave rectifier.

2. Explain how the circuit of Figure 2 can be modified to produce a pulsating DC of the opposite polarity at TP3.

3. Is the output at TP2 (a) half-wave, or (b) full-wave? _____

4. Why is R5 necessary in Figure 2?

5. What is the purpose for D2 in Figure 2?

6. Can the circuit of Figure 2 be used to rectify a 50 mV_{P-P} AC signal applied to TP1? _____

Active Limiters and Clampers

EXPERIMENT 31

PURPOSE

The purpose of this experiment is to familiarize you with active limiters and clampers. You will explore symmetrical and asymmetrical limiters, and positive and negative clampers.

FORMAL OBJECTIVES

After completing this experiment, you will be able to:

- describe the use of the op-amp in making symmetrical and asymmetrical limiters.
- use the oscilloscope and function generator to test symmetrical and asymmetrical limiters.
- describe the use of the op-amp in making positive and negative clampers.
- use the oscilloscope and function generator to test positive and negative clampers.

PREPARATIONS

In preparation for this experiment, you should study Section 12-8 of *Exploring Electronic Devices*.

COMPONENTS and EQUIPMENT

1 - Bipolar DC source (±12 V)
1 - Plug-in experiment board
1 - Function generator with leads
3 - 1N914 Diodes
1 - 10 kΩ, 1/4 W Resistor
1 - Multimeter with clip leads
1 - Dual-trace oscilloscope with leads
1 - LF356 Op-amp
1 - 100 kΩ, 1/4 W Resistor
1 - 10 μF, 35 V Electrolytic capacitor

PART ONE: ACTIVE VOLTAGE LIMITERS

In this part of the experiment, you will assemble and test the voltage limiter circuits shown in Figures 1 and 2. As you consider Figure 1, note that there are three voltage feedback paths from pin 6 to pin 2: (1) R_F serves as feedback for positive and negative alternations of the output signal, (2) D_1 serves as a feedback path for all negative alternations greater than -0.7 V, and (3) D_2 serves as a feedback path for all positive alternations greater than +0.7 V. The amplification factor for this circuit is equal to R_F/R_I for all input signals whose amplitude is below the threshold of limiting. Once limiting begins, the amplification factor is less than R_F/R_I and is determined by the conduction resistance of D_1 and D_2. Also, you should be careful to note that the output limiting level is not rock solid. In other words, the peak output voltage will increase slightly above ±0.7 V depending on how hard the amplifier is being driven.

1. Assemble the active limiter circuit of Figure 1. Apply ±12 V as indicated.
2. Use your function generator to apply a 1 kHz sine wave to the input. Start with the input signal level at minimum.
3. Use your oscilloscope to observe and compared the input and output waveforms.
4. Increase the level of the input signal until the output signal reaches limiting. This should be at an output voltage of approximately 1.4 $V_{P\text{-}P}$. This is the threshold of limiting for the output signal. Measure the actual peak-to-peak output voltage carefully and record it on the answer sheet for Part One.

Figure 1

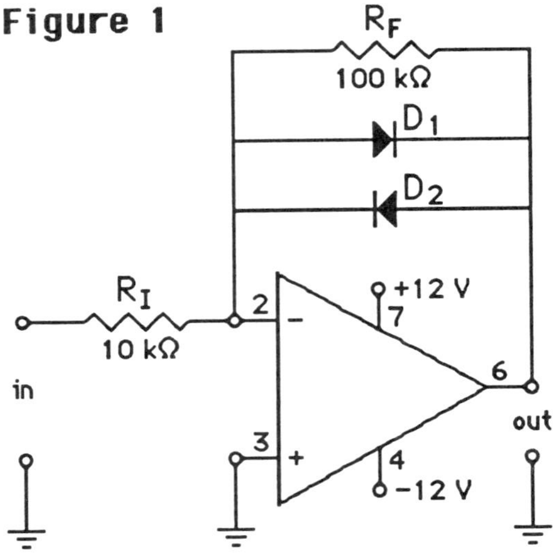

5. Measure the peak-to-peak input signal and record it on the answer sheet for Part One. This is the input threshold of limiting.
6. Use the input and output threshold voltages to determine the amplification factor of the circuit.
7. Now increase the amplitude of the input signal slowly and observe the output signal. Describe what you see on the answer sheet.
8. Determine the change in limited peak-to-peak output voltage that results from a change in input voltage from 200 $mV_{P\text{-}P}$ to 2 $V_{P\text{-}P}$.
9. Remove power. Add diode D_3 to your circuit as shown in Figure 2.
10. Apply power and an input signal to determine how the circuit is now functioning differently.

Figure 2

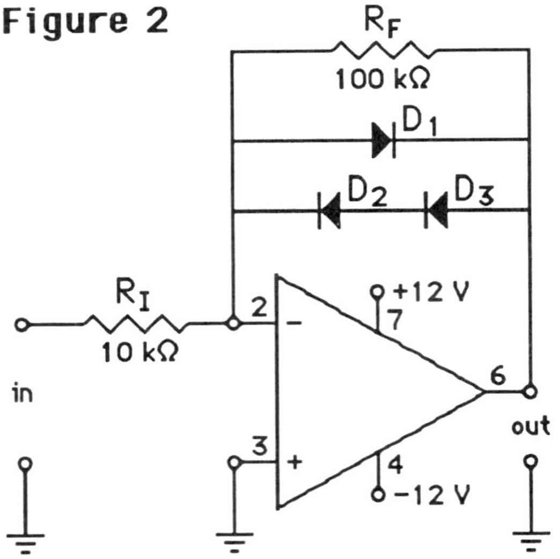

Important Note: If Channel 2 of the oscilloscope is connected to the output, momentarily set the Channel 2 input mode select switch to the GND position to establish a zero-reference line on the CRT screen. Once the reference is established the mode select switch should be in the DC position.

11. Remove power and answer all remaining questions for Part One.

PART TWO: ACTIVE CLAMPER CIRCUITS

In this part of the experiment, you will assemble and test the active clamper circuit shown in Figure 3. Recall from your reading that the active clamper is able to shift the entire waveform above or below the zero-reference line so the positive or negative peak is at exactly 0 V DC. This is an improvement over conventional diode clampers.

1. Assemble the active clamper shown in Figure 3. Apply ±12 V as shown.
2. Use your function generator to apply a 1 kHz, 5 V_{P-P} sine wave to the input.
3. Use your oscilloscope to observe and compared the input and output waveforms.

   Important Note: Momentarily set the input mode select switches to the GND position to establish a zero-reference line on the CRT screen for each channel. Once the references are established the mode select switches should be in the DC position.

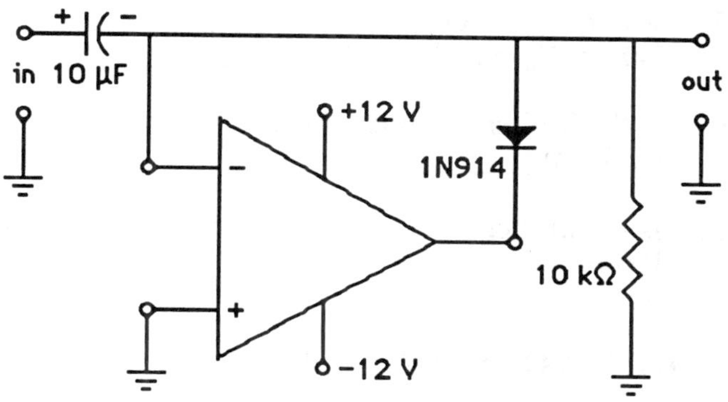

Figure 3

4. Answer all questions on the answer sheet for Part Two.
5. Remove Power.

NAME:____________________________________CLASS:__________DATE:__________

PART ONE QUESTIONS

1. What is the output voltage threshold of limiting? ____________$V_{P\text{-}P}$

2. What is the threshold of limiting for the input signal? ____________$V_{P\text{-}P}$

3. What is the amplification factor for the amplifier before limiting?
   $A_V = \mathcal{V}_{out}/\mathcal{V}_{in}$ = ____

4. Describe what happens when the input voltage is increased past the threshold of limiting.
   ____
   ____

5. When $\mathcal{V}_{in}$ = 200 $mV_{P\text{-}P}$, $\mathcal{V}_{out}$ = ____________$V_{P\text{-}P}$.
   When $\mathcal{V}_{in}$ = 2 $V_{P\text{-}P}$, $\mathcal{V}_{out}$ = ____________$V_{P\text{-}P}$.

6. Is the limiter of Figure 1 (a) symmetrical, or (b) asymmetrical? _____
   Explain: ____

7. Is the limiter of Figure 2 (a) symmetrical, or (b) asymmetrical? _____
   Explain: ____

PART TWO QUESTIONS

1. Is the circuit of Figure 3 a (a) positive clamper, or (b) a negative clamper? _____

2. How can the circuit of Figure 3 be modified to become a clamper of the opposite polarity?
   ____
   ____
   ____

3. With 5 $V_{P\text{-}P}$ applied to your clamper circuit, what is the voltage of the center-reference line for the output signal? _______ VDC

4. With 5 $V_{P\text{-}P}$ applied to your clamper circuit, what is the charge voltage across the input capacitor? _______ VDC

5. Is the op-amp continually amplifying the input signal? _______ Explain: ________________
   ____
   ____
   ____

Active Integrators and Differentiators

EXPERIMENT 32

PURPOSE

The purpose of this experiment is to familiarize you with active op-amp integrators and differentiators.

FORMAL OBJECTIVES

After completing this experiment, you will be able to:

- describe the basic op-amp integrator and op-amp differentiator.
- use the oscilloscope and function generator to test active integrators and differentiators.

PREPARATIONS

In preparation for this experiment, you should study Section 12-9 of *Exploring Electronic Devices*.

COMPONENTS and EQUIPMENT

1 - Dual-polarity DC source (±12 V)
1 - Dual-trace oscilloscope with leads
2 - LF356 Op-amps
1 - 0.01 μF Film capacitor
1 - 33 kΩ, 1/4 W Resistor
1 - 220 Ω, 1/4 W Resistor
1 - Plug-in experiment board (breadboard)
1 - Function generator with leads
1 - 1 μF Electrolytic capacitor
1 - 330 kΩ, 1/4 W Resistor
3 - 10 kΩ, 1/4 W Resistors

PART ONE: THE ACTIVE INTEGRATOR

In this part of the experiment, you will assemble and test the active integrator circuit shown in Figure 1. The circuit shown has two stages: (1) the inverting active integrator utilizing U_1, and (2) the unity-gain inverting amplifier using U_2. The result of using two inverting stages is an output that is the same polarity as the input. Note that the amplification factor (A_V) for the integrator is 10 (330 kΩ/33 kΩ = 10). Also note that the open-loop gain of the op-amp itself is 100,000 (100 dB).

Figure 1

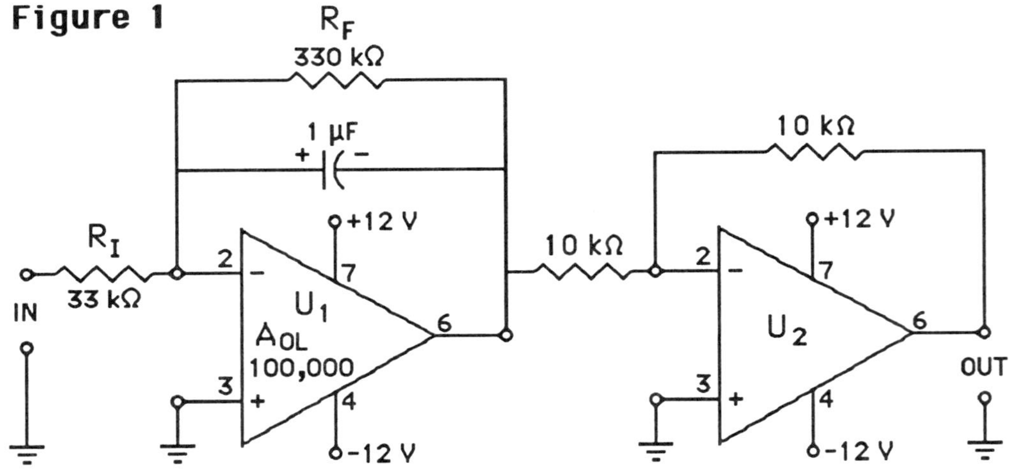

1. Assemble the circuit shown in Figure 1.

2. Use Channel 1 of your oscilloscope to monitor the input of the circuit and Channel 2 to monitor the output. The input mode select switch for each channel should be set in the DC position after using the GND position to establish a 0 V reference line for the upper and lower traces. I would suggest using the center horizontal line on the screen for the 0 V reference line for the Channel 1 upper trace and the bottom horizontal line for the Channel 2 reference line. Set the Volts/Div control for Channel 1 to 0.5 V/div. Set the Volts/Div control for Channel 2 to 2 V/div. The Time/Div control should be set at 0.5 ms/div. Trigger from Channel 1 in the Auto mode.

3. Use your function generator to apply a +1 V_P, 1 kHz pulsating DC square wave to the input. Make sure the pulsating DC square wave starts at 0 V and rises to +1 V for each pulse. You may need to adjust the DC offset control and the amplitude control several times to produce this square wave. Also, the square wave should be 50% duty cycle.

4. On the answer sheet for Part One, draw the Channel 1 and 2 signals on the screen provided.

5. Now, lower the frequency of the pulsating DC until the DC signal displayed on Channel 2 becomes rippled (noticeable triangle waveform). This should be a very low frequency.

6. Remove all power. Answer all questions for Part One.

PART TWO: THE ACTIVE DIFFERENTIATOR

In this part of the experiment, you will assemble and test the active differentiator of Figure 2. Note that U_1 is the differentiator stage and U_2 is simply a unity-gain inverter which is used to cancel the inverting effect of the differentiator stage.

Figure 2

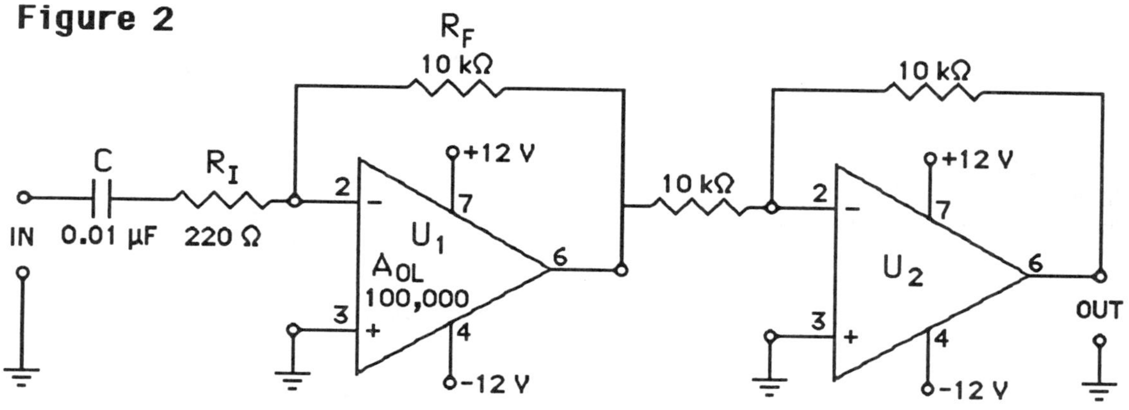

1. Assemble the circuit shown in Figure 2.
2. Use Channel 1 of your oscilloscope to monitor the input of the circuit and Channel 2 to monitor the output. The input mode select switch for each channel should be set in the DC position after using the GND position to establish a 0 V reference line for the upper and lower traces. Use the center horizontal line on the screen for the 0 V reference line for the Channel 1 upper trace and the second from the bottom horizontal line for the Channel 2 reference line. Set the Volts/Div control for Channel 1 to 50 mV/div. Set the Volts/Div control for Channel 2 to 5 V/div. The Time/Div control should be set at 100 µs/div. Trigger from Channel 1 in the Auto mode.
3. Use your function generator to apply a +100 mV_P, 4 kHz pulsating DC square wave to the input. Make sure the pulsating DC square wave starts at 0 V and rises to +100 mV for each pulse. You may need to adjust the DC offset control and the amplitude control several times to produce this square wave. Also, the square wave should be 50% duty cycle.
4. On the answer sheet for Part Two, draw the Channel 1 and 2 signals on the screen provided.
5. Now, increase the frequency of the pulsating DC and observe the output signal on the lower Channel 2 trace. Switch the oscilloscope's Time/Div control to 50 µs/div and 10 µs/div as needed as you increase the frequency of the applied DC square wave. Describe what happens to the output waveform and offer an explanation on the answer sheet.
6. Remove all power. Answer all questions for Part Two.

NAME:____CLASS:__________DATE:__________

PART ONE QUESTIONS

1. Draw the Channel 1 and 2 signals on this screen.

   Time/Div = __________ ms/div
   Channel 1 Volts/Div = __________ V/div
   Channel 2 Volts/Div = __________ V/div

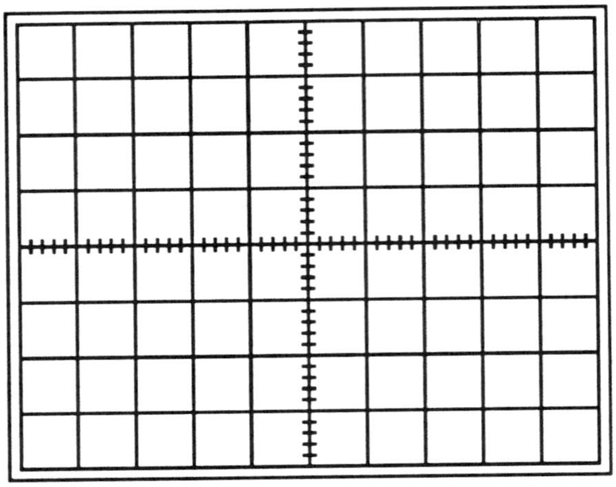

2. For the circuit of Figure 1, what does the final output voltage represent?

3. With a 1 V_P, 50% duty cycle, pulsating DC square wave applied to the input of the circuit of Figure 1, what is the voltage at pin 6 of U_1? __________ V

4. What is the voltage at the final output (pin 6 of U_2)? __________ V

5. What is the voltage amplification factor for the active integrator and how is this determined?

6. What is the average value of voltage for the pulsating DC applied to the input? __________ V

7. Show mathematically how the average input voltage is related to the final output voltage.

8. What does the percent duty cycle have to do with the average DC voltage? Demonstrate mathematically using a 50% duty cycle.

9. What is the value of Miller input capacitance for the integrator of Figure 1? Show your work.

10. What is the value of Miller input resistance for the integrator of Figure 1? Show your work.

11. What is the average DC voltage at pin 2 of the integrator? Show mathematically.

PART TWO QUESTIONS

1. Draw the Channel 1 and 2 signals on this screen.

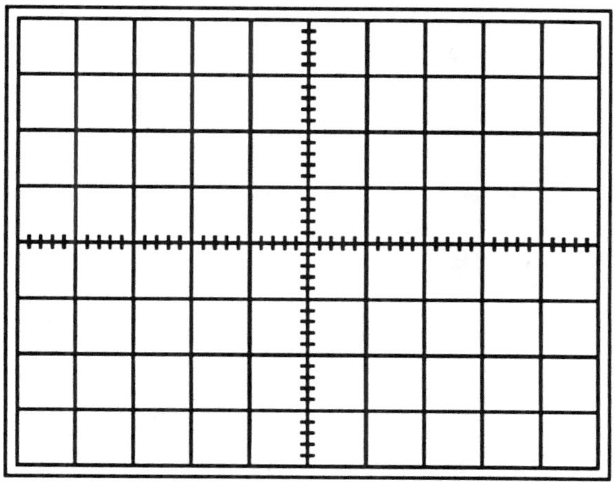

Time/Div = __________ µs/div
Channel 1 Volts/Div = __________ V/div
Channel 2 Volts/Div = __________ V/div

2. What is the amplification factor (A_V) for the differentiator stage of Figure 2? Show calculation.

3. With a +100 mV_P pulsating DC square wave applied to Figure 2, what is the peak output voltage? __________ V_P.

4. How is the peak output voltage of Figure 2 related to the peak input voltage? Show mathematically.

5. How long does it take for the input capacitor to fully charge and discharge through R_I? Demonstrate mathematically.

6. What is the pulse duration for the 4 kHz, 50% duty cycle input signal? Show your work.

7. Explain how the capacitor charge/discharge time is related to the pulse duration and the appearance of the differentiated output waveform.

8. Describe what happens to the appearance of the output waveform as the frequency of the applied pulsating DC is increased above 4 kHz.

____

____

9. As the frequency of the applied pulsating DC is increased above 4 kHz, which of these decreases?

   (a) capacitor charge/discharge time
   (b) applied pulse duration
   (c) time between output pulses

10. If we say that the input cycle time must be at least 100 times longer than 1 TC (1 TC = $R_I \cdot C$) to insure narrow output pulses, what is the highest pulse frequency that we can apply to the input of Figure 2? Show calculations.

____

____

11. To what value should C be changed to insure narrow output pulses if the applied pulse frequency is 20 kHz? Show calculations. (Hint: This is like #10 but in reverse.)

____

____

NOTES AND CALCULATIONS

Active Highpass and Lowpass Filters

EXPERIMENT 33

PURPOSE

The purpose of this experiment is to familiarize you with constant-K active highpass and lowpass filters.

FORMAL OBJECTIVES

After completing this experiment, you will be able to:

- describe the operation of active highpass and lowpass filters.
- perform the mathematical analysis of constant-K active highpass and lowpass filters.
- use the oscilloscope and function generator to perform an experimental analysis of active highpass and lowpass filters.

PREPARATIONS

In preparation for this experiment, you should study Sections 13-1 and 13-2 of *Exploring Electronic Devices.*

COMPONENTS and EQUIPMENT

1 - Dual-polarity DC source (±12 V)
1 - Dual-trace oscilloscope with leads
1 - LF356 Op-amp
1 - 10 kΩ, 1/4 W Resistor
2 - 2.2 kΩ, 1/4 W Resistors
1 - Plug-in experiment board (breadboard)
1 - Function generator with leads
2 - 0.1 μF Film capacitors
1 - 5.9 kΩ, 2%, 1/4 W Resistor
2 - 1 kΩ, 1/4 W Resistors

PART ONE: THE CONSTANT-K ACTIVE HIGHPASS FILTER

In this part of the experiment, you will assemble and test the constant-K active highpass filter shown in Figure 1. As you can see, this is a two-pole (second order) active filter having an ideal rolloff of 40 dB per decade. The K factor is equal to the amplification factor where:

$K = A_V = (R_F + R_I)/R_I$ and damping factor DF = 3 - K.

Recall from your reading that the second-order active filter produces a Bessel response if DF is greater than approximately 1.4 and a Chebyshev response if DF is less than approximately 1.4. With a DF of approximately 1.4, the filter produces the Butterworth response, also known as maximally-flat response.

1. Assemble the constant-K active highpass filter shown in Figure 1. Have someone check your wiring.

2. Connect Channel 1 of your oscilloscope to the input and Channel 2 to the output of your circuit. The input mode select switch should be set to AC for both channels. Let the Channel-1 trace occupy the upper half of the screen and the Channel-2 trace the lower half. Set the Time/Div and Volts/Div controls as needed.

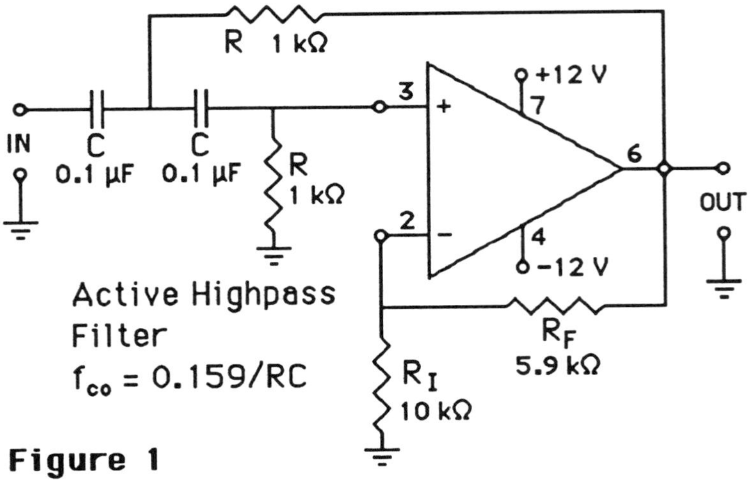

Figure 1

3. Use your function generator to apply a 2 V_{P-P}, 10 kHz sine wave to the input of your filter.

4. Determine the A_V of your filter at the 10 kHz frequency where $A_V = V_{out}/V_{in}$. Record your answer on the answer sheet for Part One.

5. Adjust the amplitude control on the function generator to set the output voltage of the active filter to 8 V_{P-P} (frequency still set at 10 kHz). The Channel 2 Volts/Div control should be set to 2 V/div so the signal occupies the entire lower half of the screen (4 cm).

6. Now, lower the frequency by adjusting the frequency-adjust knob on the function generator until the output signal occupies only 2.83 cm of the screen vertically. Measure the time for one cycle and calculate the frequency (f = 1/T). This is the cutoff frequency for the active filter - the frequency at which the signal voltage drops to $0.707 \cdot V_{max(passband)}$. Note that 2.83 cm is 0.707 • 4 cm. Record this frequency on the answer sheet for Part One.

   Note: The cutoff frequency is also called the half-power frequency where the output voltage and power is 3 dB down from maximum.

   $$\#dB = 20 \log_{10}(V_{out}/V_{max(passband)}) = 20 \log_{10}(2.83\ cm/4\ cm) = 20 \log_{10}(0.707) = -3\ dB$$

7. Use the formula $f_{co} = 0.159/RC$ to calculate the cutoff frequency of your highpass filter using the values for C and R shown in Figure 1. This should be close to the measured frequency. However, expect some error due to component tolerance, oscilloscope calibration, and reading accuracy.

8. For the following procedure, observe the input signal on Channel 1 as you change frequency to make sure the input signal does not change in amplitude. The input should remain as set in Step 5 previously.

9. Record the output voltage for every frequency listed in Table 1 on the answer sheet and convert each to decibels where: $\#dB = 20 \log_{10}(V_{out}/8\ V_{P\text{-}P})$

10. Use the data collected in Table 1 to create a dot-to-dot graph on Chart 1.

11. Remove all power and answer any remaining questions for Part One.

PART TWO: THE CONSTANT-K ACTIVE LOWPASS FILTER

In this part of the experiment, you will assemble and test the constant-K active lowpass filter shown in Figure 2. As you can see, this is a two-pole (second order) active filter having an ideal rolloff of 40 dB per decade.

1. Assemble the constant-K active lowpass filter shown in Figure 2. Have someone check your wiring.

2. Connect Channel 1 of your oscilloscope to the input and Channel 2 to the output of your circuit. The input mode select switch should be set to AC for both channels. Let the Channel-1 trace occupy the upper half of the screen and the Channel-2 trace the lower half. Set the Time/Div and Volts/Div controls as needed.

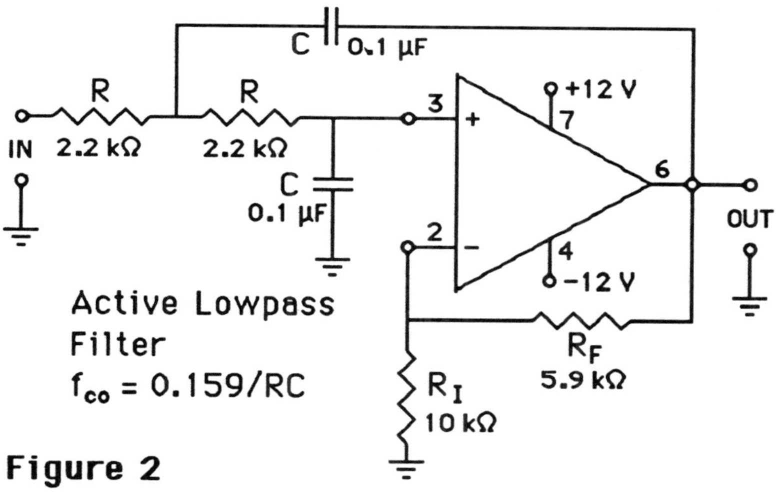

Figure 2

3. Use your function generator to apply a 2 $V_{P\text{-}P}$, 100 Hz sine wave to the input of your filter.

4. Determine the A_V of your filter at the 100 Hz frequency where $A_V = V_{out}/V_{in}$. Record your answer on the answer sheet for Part Two.

5. Adjust the amplitude control on the function generator to set the output voltage of the active filter to 8 $V_{P\text{-}P}$ (frequency still set at 100 Hz). The Channel 2 Volts/Div control should be set to 2 V/div so the signal occupies the entire lower half of the screen (4 cm).

6. Now, raise the frequency by adjusting the frequency-adjust knob on the function generator until the output signal occupies only 2.83 cm of the screen vertically. Measure the time for one cycle and calculate the frequency ($f = 1/T$). This is the cutoff frequency for the active filter - the frequency at which the signal voltage drops to $0.707 \bullet \mathcal{V}_{max(passband)}$. Note that 2.83 cm is 0.707 • 4 cm. Record this frequency on the answer sheet for Part Two.

7. Use the formula $f_{co} = 0.159/RC$ to calculate the cutoff frequency of your lowpass filter using the values for C and R shown in Figure 2. This should be close to the measured frequency. However, expect some error due to component tolerance, oscilloscope calibration, and reading accuracy.

8. For the following procedure, observe the input signal on Channel 1 as you change frequency to make sure the input signal does not change in amplitude. The input should remain as set in Step 5 previously.

9. Record the output voltage for every frequency listed in Table 2 on the answer sheet and convert each to decibels where: $\#dB = 20 \log_{10}(\mathcal{V}_{out}/8\ V_{P\text{-}P})$

10. Use the data collected in Table 2 to create a dot-to-dot graph on Chart 2.

11. Remove all power and answer any remaining questions for Part Two.

NAME:________________ CLASS:________ DATE:________

PART ONE QUESTIONS

1. $A_{V(10\ kHz)} = V_{out}/2\ V_{P\text{-}P} =$ ______________

2. $T_{fco} =$ ______________, $f_{co} = 1/T =$ ______________ Hz

3. $f_{co} = 0.159/RC =$ ______________________________ Hz

Table 1

Frequency	$V_{P\text{-}P(out)}$	#dB
10 kHz	8	0
5 kHz	_____	_____
1 kHz	_____	_____
500 Hz	_____	_____
100 Hz	_____	_____

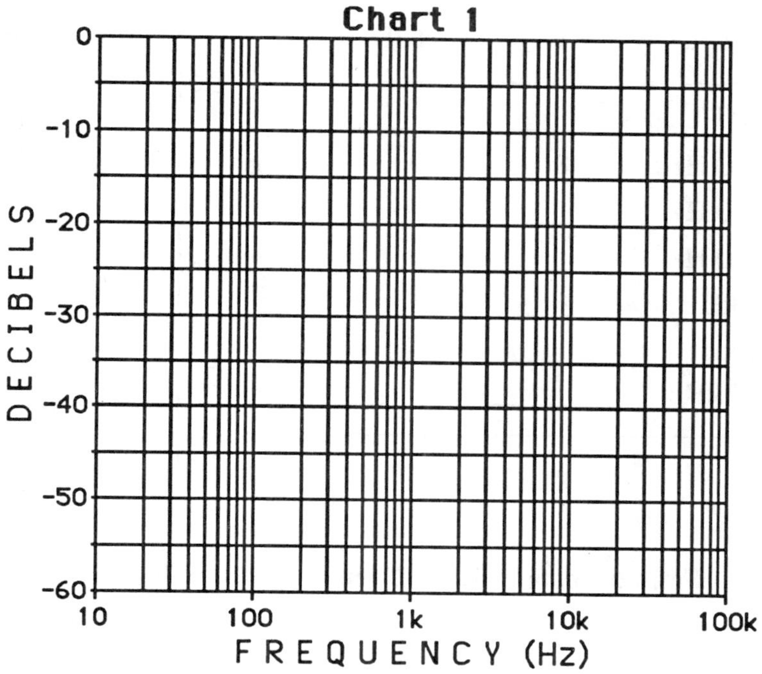

4. From the data in Table 1 determine the rolloff in dB/octave from 1 kHz to 500 Hz.
____

5. From the data in Table 1 determine the rolloff in dB/decade from 1 kHz to 100 Hz.
____

6. How closely do your calculations for slope of rolloff match the ideal?
____

7. From the graph of Chart 1, would you say the frequency response is closer to (a) Bessel, (b) Butterworth, or (c) Chebyshev? _____

PART TWO QUESTIONS

1. $A_{V(100\,Hz)} = V_{out}/2\ V_{P\text{-}P}$ = ______________

2. T_{fco} = ______________, $f_{co} = 1/T$ = ______________ Hz

3. $f_{co} = 0.159/RC$ = ______________________________ Hz

Table 2

Frequency	$V_{P\text{-}P(out)}$	#dB
50 Hz	_____	_____
100 Hz	8	0
250 Hz	_____	_____
500 Hz	_____	_____
1 kHz	_____	_____
5 kHz	_____	_____
10 kHz	_____	_____

Chart 2

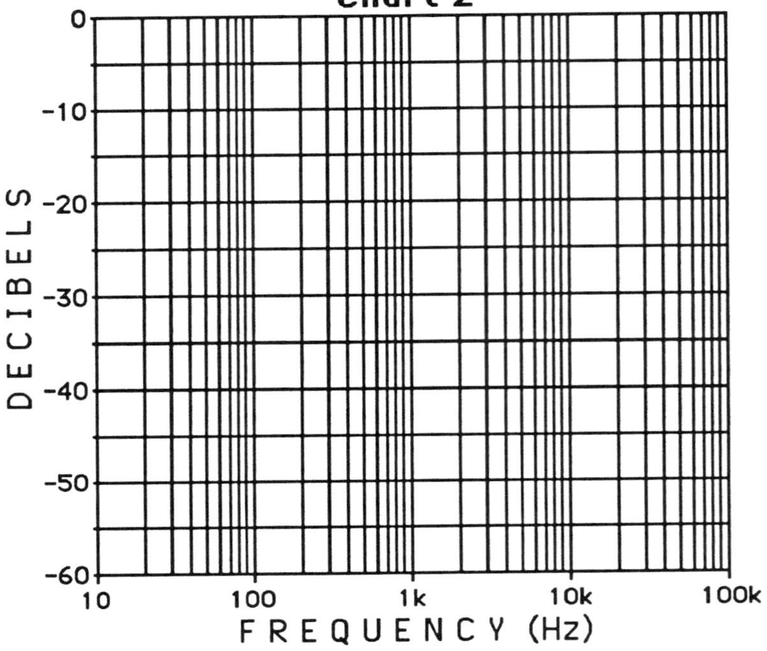

4. From the data in Table 2 determine the rolloff in dB/octave from 500 Hz to 1 kHz.

____

5. From the data in Table 2 determine the rolloff in dB/decade from 1 kHz to 10 kHz.

____

6. How closely do your calculations for slope of rolloff match the ideal?

____

7. From the graph of Chart 2, would you say the frequency response is closer to (a) Bessel, (b) Butterworth, or (c) Chebyshev? _____

The State-Variable Filter

EXPERIMENT 34

PURPOSE

The purpose of this experiment is to familiarize you with the state-variable active filter. You will assemble and test such a filter to verify circuit theory and operation.

FORMAL OBJECTIVES

After completing this experiment, you will be able to:

- identify the state-variable active filter.
- mathematically analyze the state-variable active filter.
- experimentally analyze the state-variable active filter using the oscilloscope and function generator.

PREPARATIONS

In preparation for this experiment, you should study Sections 13-3 and 13-4 of *Exploring Electronic Devices*.

COMPONENTS and EQUIPMENT

1 - Dual-polarity DC source (±12 V)
1 - Dual-trace oscilloscope with leads
3 - LF356 Op-amps
4 - 10 kΩ, 1/4 W Resistors
1 - 1 kΩ, 1/4 W Resistor
1 - Plug-in experiment board (breadboard)
1 - Function generator with leads
2 - 0.01 μF Film capacitors
2 - 4.7 kΩ, 1/4 W Resistors

DISCUSSION

In the following steps, you will assemble the state-variable active filter shown in Figure 1. You will use your oscilloscope and function generator to determine the frequency response of the various outputs. Be sure to review the operation of this filter in Section 13-3 of *Exploring Electronic Devices*. Also make sure you carefully study Section 13-4. After completing this experiment, you may wish to replace R and C values to design a filter for a frequency you desire. Also, you may want to make R_A or R_B variable so you can adjust the Q of the filter.

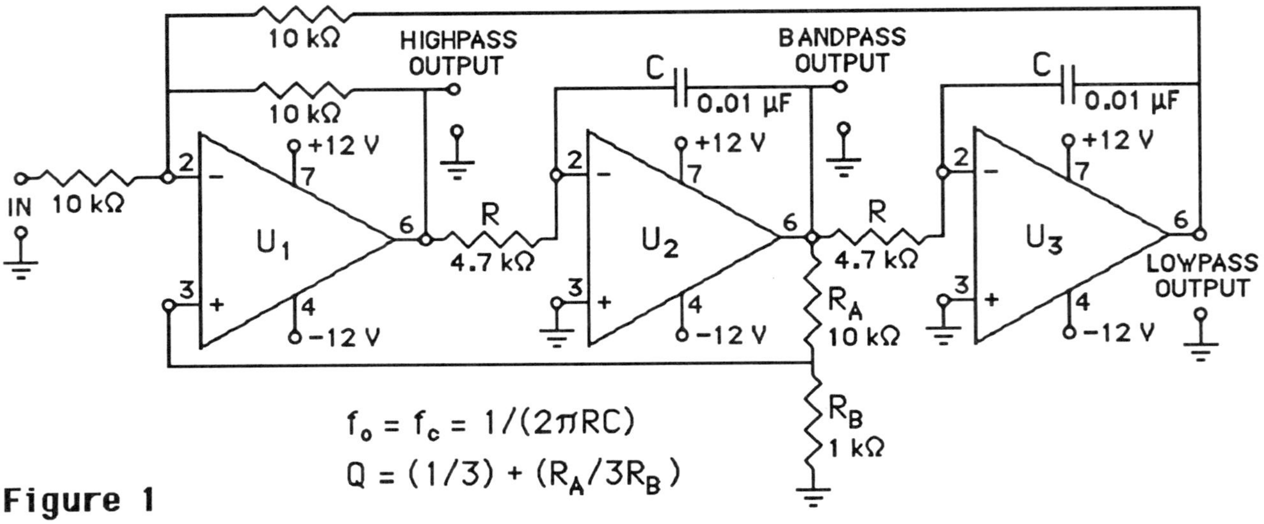

Figure 1

PROCEDURE

1. Carefully construct the circuit shown in Figure 1. Have someone check your wiring before you continue.

2. From Figure 1, determine the theoretical center frequency, Q, and bandwidth for this filter. Record your calculations on the answer sheet.

3. Use the function generator and oscilloscope to determine the highpass cutoff frequency at the highpass output, the lowpass cutoff frequency at the lowpass output, and the center frequency at the bandpass output.

   Connect the function generator to the input . The generator should be set to produce a sine-wave signal. When testing an output, adjust the frequency-adjust knob and range switches on the function generator as needed to obtain a maximum passband output signal far from the cutoff frequency. For example, when testing the lowpass output, the output signal will be at a maximum unattenuated level below 1 kHz. Thus, you may want to set the generator frequency for 500 Hz. At this frequency, the function generator's amplitude control can be adjusted to establish a reference level at the lowpass filter output. This reference level can be 8 $V_{P\text{-}P}$ to fill the oscilloscope screen with the Volts/Div control set at 1 V/div. Thus, the 500 Hz, 8 $V_{P\text{-}P}$ output signal becomes the reference signal considered to be at 0 dB. As the generator frequency is slowly increased, the output voltage will begin to drop down below

8 V_{P-P} until the cutoff frequency is reached where the output voltage equals 0.707 • 8 V_{P-P} = 5.66 V_{P-P} = -3 dB. A suggested highpass reference frequency might be 10 kHz.

For the bandpass output, the center frequency will be the reference frequency where the output voltage is set to 8 V_{P-P}. Upper and lower cutoff frequencies can then be determined.

4. Test the filter over a wide range of frequencies and plot the highpass, lowpass, and bandpass responses all on Chart 1. Use a different color writing instrument for each response curve. Label all key frequencies. The decibel level at each frequency will be:

   $\#dB = 20 \log_{10}(V_{out}/V_{ref})$, for example, $\#dB = 20 \log_{10}(V_{out}/8\ V_{P-P})$

   At a cutoff frequency, $\#dB = 20 \log_{10}(5.66\ V_{P-P}/8\ V_{P-P}) = 20 \log_{10}(0.707) = -3$ dB

   Measure the output voltage for at least 10 different test frequencies for each filter output. Create your own table of data. Convert each voltage to dBs so it can be plotted on the Bode chart (Chart 1).

5. Determine the bandpass bandwidth and circuit Q where: $BW = Uf_{co} - Lf_{co}$, and $Q = f_c/BW$.

6. Remove all power and answer any remaining questions.

NAME:______________________________ CLASS:__________ DATE:__________

QUESTIONS

1. Calculated Values: $f_c = 1/(2\pi RC) =$ ____________________
   $Q = (1/3) + (R_A/3R_B) =$ ____________________
2. Use the calculated values above to determine the theoretical bandwidth.
   $BW = f_c/Q =$ ____________________
3. Using Test Equipment: $Uf_{co} =$ __________ Hz, $Lf_{co} =$ __________ Hz
   $f_c =$ __________ Hz, $BW =$ __________ Hz
4. Actual $Q = f_c/BW =$ ____________________
5. How can the Q of this circuit be increased?
   ____
6. How can the center frequency of this circuit be lowered?
   ____

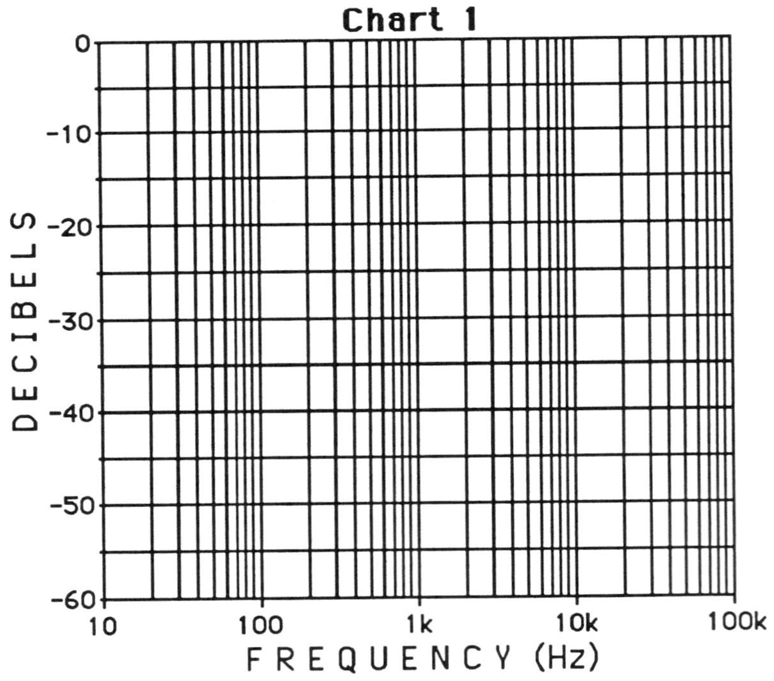

Linear Op-Amp Mixers

EXPERIMENT 35

PURPOSE

The purpose of this experiment is to familiarize you with a basic op-amp audio mixer. This is a linear mixer acting as a summing amplifier.

FORMAL OBJECTIVES

After completing this experiment, you will be able to:

- identify linear op-amp mixers.
- predict maximum output voltage when input voltages and input weights are known.
- use the function generator and oscilloscope to analyze a linear op-amp mixer.

PREPARATIONS

In preparation for this experiment, you should study Section 14-1 of *Exploring Electronic Devices*.

COMPONENTS and EQUIPMENT

1 - Dual-polarity DC source (±12 V)
1 - Dual-trace oscilloscope with leads
1 - LF356 Op-amp
1 - 33 kΩ, 1/4 W Resistor
1 - Plug-in experiment board (breadboard)
2 - Function generators with leads
1 - 100 kΩ, 1/4 W Resistor
1 - 56 kΩ, 1/4 W Resistor

DISCUSSION

In the following steps, you will analyze, assemble, and test the linear mixer circuit shown in Figure 1. The emphasis here will be on the two weighted inputs and the linearly combined output that results from the two AC inputs.

Figure 1

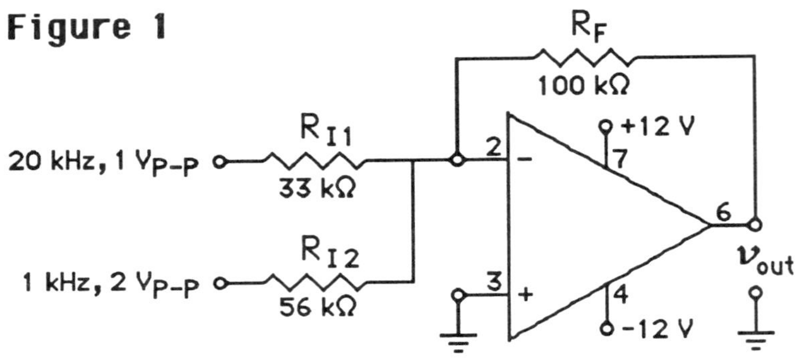

PROCEDURE

1. Before assembling the circuit of Figure 1, take a few moments to determine the weight of each input where the weight of Input 1 = $K_1 = A_{V1} = R_F/R_{I1}$ and the weight of Input 2 = $K_2 = A_{V2} = R_F/R_{I2}$. Use the weights to calculate the peak-to-peak output voltage resulting from each individual input and the combined peak-to-peak output voltage. Record your answers on the answer sheet.

2. Construct the circuit shown in Figure 1. Use an LF356 op-amp. Apply ±12 V DC as indicated.

3. With no signal applied to Input 2, apply a 20 kHz, 1 $V_{P\text{-}P}$ sine wave to Input 1 of your circuit and measure the resulting peak-to-peak output voltage. Record your reading on the answer sheet and use it to determine the weight of Input 1 where $K_1 = A_{V1} = v_{out}/v_{in}$.

4. With no signal applied to Input 1, apply a 1 kHz, 2 $V_{P\text{-}P}$ sine wave to Input 2 of your circuit and measure the resulting peak-to-peak output voltage. Record your reading on the answer sheet and use it to determine the weight of Input 2 where $K_2 = A_{V2} = v_{out}/v_{in}$.

5. Use two function generators so you can apply the 1 kHz and 20 kHz signals at the same time as shown in Figure 1. Observe the output with your oscilloscope and determine the combined peak-to-peak voltage. The Time/Div control should be set to 0.2 ms/div (200 μs/div). The Volts/Div control should be set to 1 or 2 V/div.

6. Remove power and answer all remaining questions on the answer sheet.

   Extra: You may also want to experiment with mixing DC on one input with AC on the other or mixing different DC voltages using the two inputs.

NAME:________________________________ CLASS:__________ DATE:__________

QUESTIONS

1. $K_1 = A_{V1} = R_F/R_{I1} =$ ______________________________

2. $K_2 = A_{V2} = R_F/R_{I2} =$ ______________________________

3. $\mathcal{V}_{out(Input\ 1)} = 1\ V_{P\text{-}P} \bullet K_1 =$ ______________________________

4. $\mathcal{V}_{out(Input\ 2)} = 2\ V_{P\text{-}P} \bullet K_2 =$ ______________________________

5. $\mathcal{V}_{out(Combined)} = (1\ V_{P\text{-}P} \bullet K_1) + (2\ V_{P\text{-}P} \bullet K_2) =$ ______________________________

6. Measured Value: $\mathcal{V}_{out(Input\ 1)} =$ ________ $V_{P\text{-}P}$, $K_1 = A_{V1} = \mathcal{V}_{out}/\mathcal{V}_{in} =$ ______________

7. Measured Value: $\mathcal{V}_{out(Input\ 2)} =$ ________ $V_{P\text{-}P}$, $K_2 = A_{V2} = \mathcal{V}_{out}/\mathcal{V}_{in} =$ ______________

8. Measured Value: $\mathcal{V}_{out(Combined)} =$ ________ $V_{P\text{-}P}$

9. Write a brief description of the combined output signal.

____

____

10. Were any new frequencies created when the two input signals were mixed to create the output signal? ________ Why?

____

____

NOTES AND CALCULATIONS

The 555 Timer IC

EXPERIMENT 36

PURPOSE

The purpose of this experiment is to familiarize you with the 555 timer IC. In this experiment, you will implement the 555 first as a timer (one-shot, monostable multivibrator), then as a square-wave generator (astable multivibrator).

FORMAL OBJECTIVES

After completing this experiment, you will be able to:

- describe the purpose for each pin of the 555 IC.
- explain the operation of the 555 timer circuit.
- explain the operation of the 555 square-wave generator.
- use the oscilloscope to verify the operation of 555 circuits.

PREPARATIONS

In preparation for this experiment, you should study Section 15-4 of *Exploring Electronic Devices*.

COMPONENTS and EQUIPMENT

1 - DC source (+5 V)
1 - Plug-in experiment board (breadboard)
1 - LM555 Timer IC
1 - 10 μF Electrolytic capacitor
1 - 0.01 μF Film capacitor
2 - 10 kΩ, 1/4 W Resistors
1 - Digital multimeter with clip leads
1 - Dual-trace oscilloscope with leads
1 - 1N914 or 1N4148 Diode
1 - 0.1 μF Film capacitor
1 - 1 MΩ, 1/4 W Resistor

PART ONE: THE 555 MONOSTABLE MULTIVIBRATOR - ONE-SHOT TIMER

In this part of the experiment, you will assemble and test the 555 timer circuit shown in Figure 1. Recall from your reading that pin 2 is the control pin and pin 3 is the pulse output pin. When pin 2 is momentarily brought low, pin 3 goes high and the timing capacitor C begins to charge through timing resistor R. The length of the output pulse is T = 1.1RC. When the capacitor voltage reaches 2/3 of the supply voltage, as sensed by pin 6, pin 7 goes low and quickly discharges the capacitor. Thus, pin 6 is the threshold sense pin and pin 7 is the discharge pin. When the capacitor is discharged, the output pulse at pin 3 goes low, which is its normal condition until pin 2 goes low again. Pin 5 is the filter pin where the 0.01 μF capacitor acts as a filter to insure a steady DC reference voltage for the internal comparator that is connected internally to pin 6. Pin 1 is the ground pin and pin 8 is the DC source voltage pin. Pin 4 is a master reset control pin that can be used to reset pin 3 to 0 V any time during an output timing pulse. For most applications, pin 4 is tied to the DC source voltage to insure that it remains disabled.

Figure 1

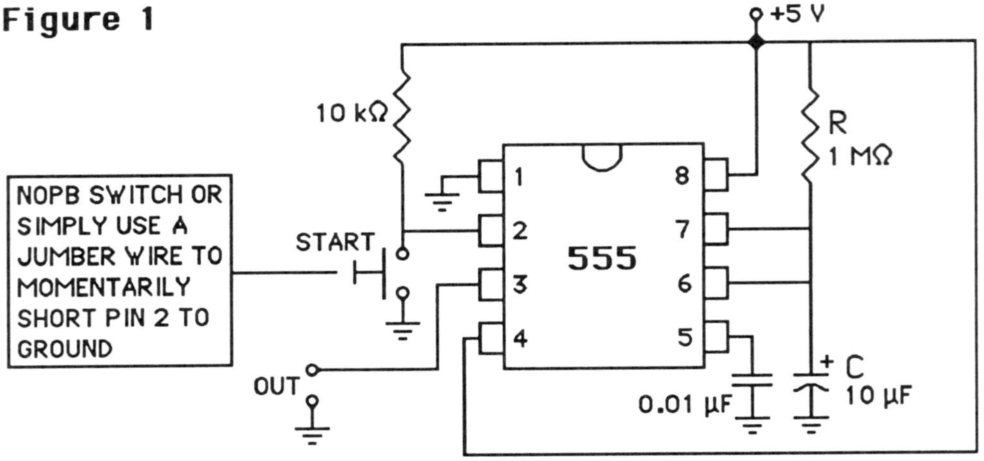

1. Before assembling and testing the timer circuit of Figure 1, calculate the length of the output timing pulse and record your answer on the answer sheet for Part One. T = 1.1RC
2. Assemble the 555 timer circuit of Figure 1. Have someone check your work.
3. Apply +5 V as shown.
4. Connect your oscilloscope to the output, pin 3. Set the input mode select on DC after establishing a 0 reference line on the screen. The Volts/Div control can be set to 1 V/div.
5. Momentarily short pin 2 to ground to start the timing pulse. Use a watch or clock to time the output pulse and compare it to your calculated value.
6. Remove power and answer all questions for Part One.

PART TWO: THE 555 ASTABLE MULTIVIBRATOR

In this part of the experiment, you will configure the 555 as an astable multivibrator as shown in Figure 2. In this configuration, the 555 acts as a continuous square-wave pulse generator. The output pulse timing operation is automatically restarted when the capacitor is discharged slightly below 1/3 VDC. In other words, the output pulse at pin 3 goes high when the capacitor voltage at pin 2 drops slightly below 1/3 VDC. When the timing capacitor voltage reaches slightly more than 2/3 VDC, as sensed by pin 6, the output pulse drops back to 0 V and pin 7 goes low to discharge the capacitor. Note that C_T charges through $R_{T1} + R_{T2}$ and discharges only through RT2. This causes the percent duty cycle to be greater than 50%. If a near 50% duty cycle is needed, a small diode can be placed across R_{T2} as suggested in Figure 2. This diode acts as a current bypass so timing capacitor C_T charges through only R_{T1} via the diode. Thus, the charge and discharge resistances are equal creating a 50% duty cycle.

Pulse ON time = $t_1 = 0.693 \bullet C_T \bullet (R_{T1} + R_{T2})$ when the bypass diode is *not* used, and $t_1 = 0.693 \bullet C_T \bullet R_{T1}$ when the bypass diode is used.

Pulse OFF time = $t_2 = 0.693 \bullet C_T \bullet R_{T2}$ whether the diode is used or not.

The pulse repetition rate, or frequency, of the square wave pulses is equal to 1/T where:

$T = t_1 + t_2$

$T = 0.693 \bullet C_T \bullet (R_{T1} + 2R_{T2})$ when the bypass diode is *not* used, and

$T = 0.693 \bullet C_T \bullet (R_{T1} + R_{T2})$ when the bypass diode is used.

Figure 2

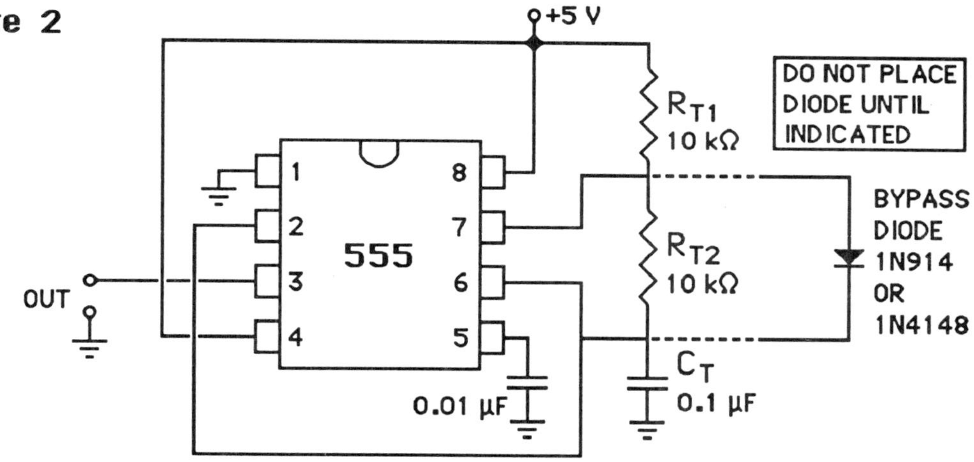

1. Before assembling the circuit of Figure 2, calculate the pulse ON time, pulse OFF time, total cycle time, and the pulse repetition rate, or frequency, *without* the bypass diode. Record your work on the answer sheet for Part Two.

2. Calculate the pulse ON time, pulse OFF time, total cycle time, and the pulse repetition rate, or frequency, *with* the bypass diode. Record your work on the answer sheet for Part Two.

3. Assemble the astable multivibrator circuit of Figure 2. Do not place the diode across R_{T2} at this time. Have someone check your work.

4. Apply +5 V as shown.

5. Use Channel 1 of your oscilloscope to observe the output pulses at pin 3 and use Channel 2 to observe the timing capacitor voltage.

6. From the oscilloscope display, determine the pulse ON time t_1, pulse OFF time t_2, and total cycle time T. Then calculate the pulse frequency (f = 1/T). Record your findings on the answer sheet for Part Two.

7. Use the ON and OFF times to determine the percent duty cycle where:
   % Duty cycle = 100% • t_1/T

8. Place the diode across R_{T2} as suggested in Figure 2.

9. From the oscilloscope display, determine the pulse ON time t_1, pulse OFF time t_2, and total cycle time T. Then calculate the pulse frequency (f = 1/T). Record your findings on the answer sheet for Part Two.

10. Answer all remaining questions for Part Two.

NAME:________________________________CLASS:__________DATE:__________

PART ONE QUESTIONS

1. Calculated timing pulse: $T = 1.1RC =$ ______________________
2. Measured timing pulse: ___________
3. How can the length of the timing pulse be increased? ______________________
4. How can the length of the timing pulse be decreased? ______________________
5. What is the length of the timing pulse if C is changed to 1 μF? ______________
6. What is the purpose for pin 6 of the 555 IC? ______________________
____
7. Why is pin 4 tied to the DC source voltage? ______________________
____
8. Is the capacitor (a) charging or is the capacitor (b) discharging while the output pin is high? __

PART TWO QUESTIONS

1. *Without* the bypass diode: (theoretical values)
   $t_1 =$ ______________________
   $t_2 =$ ______________________
   $T =$ ______________________
   $f = 1/T =$ ______________________
2. *With* the bypass diode: (theoretical values)
   $t_1 =$ ______________________
   $t_2 =$ ______________________
   $T =$ ______________________
   $f = 1/T =$ ______________________
3. *Without* the bypass diode: (using measured values)
   $t_1 =$ ______________________
   $t_2 =$ ______________________
   $T =$ ______________________
   $f = 1/T =$ ______________________
4. *With* the bypass diode: (using measured values)
   $t_1 =$ ______________________
   $t_2 =$ ______________________
   $T =$ ______________________
   $f = 1/T =$ ______________________

5. From your Channel 1 and Channel 2 observations, is the timing capacitor (a) charging, or (b) discharging while the output pulse is high? ____

6. What is the purpose for the bypass diode?

7. What effects does the bypass diode have on circuit operation?

8. How can the pulse repetition rate, or frequency, be changed and still maintain a 50% duty cycle? Suggest two methods.

9. **Challenge Question:** How can the circuit of Figure 2 have a *near* 50% duty cycle without using the bypass diode? Suggest timing resistor values to accomplish this where the pulse frequency is approximately 1 kHz and $C_T = 0.01\ \mu F$. Show all work.

NOTES AND CALCULATIONS

The XR2206 Function Generator

EXPERIMENT **37**

PURPOSE

The purpose of this experiment is to familiarize you with the XR2206 function generator IC. This is a very versatile integrated circuit that is often used as a laboratory function generator, low-frequency FM modulator, and low-frequency AM modulator. In this experiment, the XR2206 will be implemented only as a sine/triangle/square-wave function generator. This circuit can be utilized as a valuable test instrument with the addition of a power supply, cabinet, calibrated dial for frequency adjust, amplitude control, function switch, and range switch to switch in different values of timing capacitors.

FORMAL OBJECTIVES

After completing this experiment, you will be able to:

- identify the XR2206 implemented as a function generator.
- utilize the XR2206 as a multirange function generator.
- use the oscilloscope to test the XR2206 IC.

PREPARATIONS

In preparation for this experiment, you should study Section 15-3 of *Exploring Electronic Devices*.

COMPONENTS and EQUIPMENT

1 - DC Power supply (+12 V)
1 - Oscilloscope with test leads
2 - 10 μF Electrolytic capacitors
1 - 0.1 μF Film capacitor
1 - 100 kΩ Trimmer potentiometer
1 - 1 kΩ Trimmer potentiometer
2 - 4.7 kΩ, 1/4 W Resistors
1 - Plug-in experiment board (breadboard)
1 - XR2206 Function generator IC
2 - 1 μF Electrolytic capacitors
1 - 0.01 μF Film capacitor
1 - 10 kΩ Trimmer potentiometer
1 - 10 kΩ, 1/4 W Resistor
1 - 1 kΩ, 1/4 W Resistor

DISCUSSION

In the following steps, you will assemble and test the function generator shown in Figure 1. You will find the ZR2206 to be very easy to use or design around. Notice from Figure 1 there are two main controls, the 100 kΩ amplitude control and the 10 kΩ frequency adjust. The 1 kΩ shape trim adjust control is adjusted only once to obtain a nearly perfect sine wave output. Note also the jumper wire at pin 14 which is used to select the sine-wave or triangle-wave output at pin 2.

If you decide to build your own shop function generator from this experiment, you will want to use a SPST switch instead of the jumper. Also, you will want to use full-size panel-mount potentiometers for the amplitude control and the frequency adjust. To provide frequency range selection, you will want to use a rotary switch to switch in different values of timing capacitors between pins 5 and 6.

Figure 1

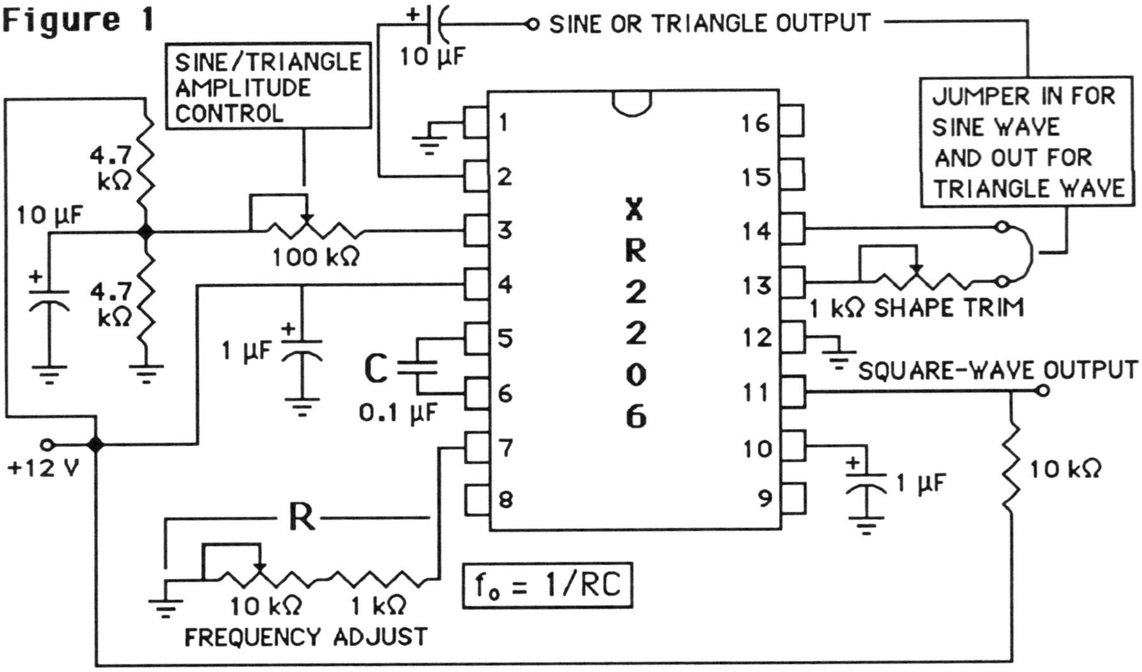

PROCEDURE

1. Using the formula $f_o = 1/RC$, determine the range of frequency adjustment for the circuit of Figure 1. Show all work on the answer sheet.

2. Assemble the circuit of Figure 1.

   Important: Have someone carefully check your work before applying power.

3. Connect Channel 1 of your oscilloscope to the square-wave output, pin 11.

4. Connect Channel 2 of your oscilloscope to the sine/triangle output, the 10 µF capacitor at pin 2.

5. Apply +12 VDC to your circuit as shown.

6. Set the oscilloscope controls as needed to place the square wave in the upper half of the screen and the sine/triangle wave in the lower half of the screen.

7. Determine if the square wave is a pulsating DC or is an AC square wave.

8. Remove the jumper wire and observe the triangle-wave output. Replace the jumper wire.

9. Set the amplitude control (100 kΩ trimmer) midway.

10. Set the frequency adjust control (10 kΩ trimmer) midway.

11. With the jumper wire in place, adjust the shape trimmer potentiometer to obtain a well-shaped sine-wave output.

    Note: Without the use of an audio distortion analyzer, the shape adjustment is purely subjective. In other words, you will adjust the shape until you think it looks like an accurate sine wave.

12. Adjust the amplitude control over its full range and determine the minimum and maximum peak-to-peak sine-wave voltages at the output. Record your findings on the answer sheet.

13. Adjust the frequency control over its full range and determine the minimum and maximum output frequencies. Record your findings on the answer sheet.

14. Replace the 0.1 μF timing capacitor with a 0.01 μF capacitor and check the frequency range again. Record your findings on the answer sheet.

15. Remove power and answer all remaining questions on the answer sheet.

NAME:______________________________ CLASS:__________ DATE:__________

QUESTIONS

1. Calculated frequency range:
   $f_{o(low)} = 1/RC =$ ______________________________
   $f_{o(high)} = 1/RC =$ ______________________________

2. Is the square wave signal at pin 11 a (a) pulsating DC square wave or an (b) AC square wave?

3. Is the triangle-wave signal (a) a well-shaped triangle wave or is it (b) more of a sawtooth wave?

4. What are the minimum and maximum sine-wave output voltages?
   $V_{out(min)} =$ __________ $V_{P\text{-}P}$ $V_{out(max)} =$ __________ $V_{P\text{-}P}$

5. With the 0.1 μF timing capacitor, what is the frequency range?
   $f_{o(min)} =$ __________ Hz to $f_{o(max)} =$ __________ Hz

6. With the 0.01 μF timing capacitor, what is the frequency range?
   $f_{o(min)} =$ __________ Hz to $f_{o(max)} =$ __________ Hz

7. What is the ratio of frequency range adjustment? __________

8. List four values of timing capacitors needed to cover an approximate frequency range of 20 Hz to 200 kHz in 4 ranges. Assume the same R values are used, 10 kΩ potentiometer with 1 kΩ fixed series resistor. Show all calculations here.

   ____
   ____
   ____
   ____
   ____
   ____
   ____

An SCR Circuit

EXPERIMENT 38

PURPOSE

The purpose of this experiment is to familiarize you with the operation of the silicon controlled rectifier (SCR). Here, you will build and test an SCR control circuit.

FORMAL OBJECTIVES

After completing this experiment, you will be able to:

- recognize an SCR control circuit.
- use the oscilloscope to verify the operation of the SCR circuit.
- explain the operation of the SCR in terms of delay angle and conduction angle.

PREPARATIONS

In preparation for this experiment, you should study Section 16-2 of *Exploring Electronic Devices*. Also, you will need to solder two wire leads to your lamp so it can be connect in your circuit. A bayonet socket can be used, but leads will need to be soldered to the socket terminals.

COMPONENTS and EQUIPMENT

1 - Dual-trace oscilloscope with leads
1 - 25.2 V CT, 450 mA Transformer
1 - 2N1595/6 SCR, 50/100 V, 1.6 A max
1 - 10 kΩ Trimmer potentiometer
1 - 470 Ω, 1/4 W Resistor
1 - plug-in experiment board (breadboard)
1 - #1815 Lamp, 14 V, 0.2 A
2 - 1N4001 Diodes, 50 V, 1 A
1 - 1 kΩ, 1/4 W Resistor

DISCUSSION

In the following steps, you will assemble and test the SCR circuit of Figure 1. As you can see, full wave rectification is used at the secondary of the transformer. Therefore, the peak pulsating DC voltage will be approximately ((25.2 V / 2) • 1.414) - 0.7 V = 17.1 V_P. The average DC voltage will be approximately 17.1 V_P • 0.636 = 10.9 VDC. This average DC voltage is below the 14 V rating of the #1815 lamp.

Recall from your reading that the SCR will remain off until it is triggered on. The anode-to-cathode voltage (V_{AK}) at which the SCR triggers is determined by the amount of gate current (I_G). As you can see in Figure 1, the gate current can be adjusted with the 10-kΩ potentiometer, wired as a rheostat. Thus, the delay angle and angle of conduction for the SCR can be varied. In this experiment, you will note that, when the SCR is conducting, its conduction angle can only be varied over a range of 90° to 180°. Thus, the SCR offers no low-level brightness control of the lamp.

The SCR used in this experiment should be either the 2N1595 (V_{DRM} = V_{RRM} = 50 V) or the 2N1596 (V_{DRM} = V_{RRM} = 100 V), or an equivalent replacement. The maximum RMS on-state current ($I_{T(RMS)}$) for these devices is 1.6 A. The gate trigger current (I_{GT}) is typically 2 mA (rated with a 12 Ω load resistance and 7 V applied). The holding current (I_H) is typically 5 mA.

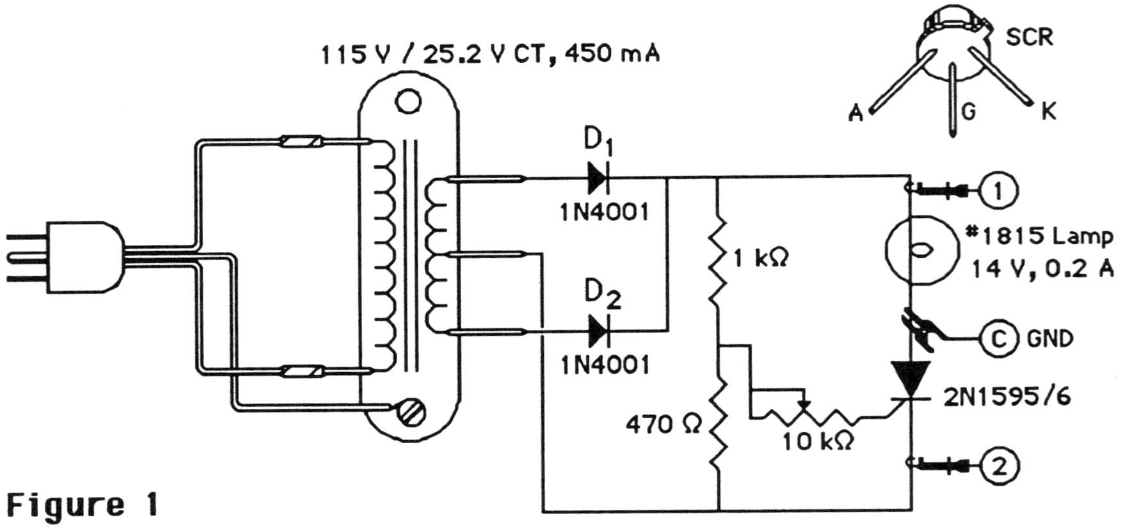

Figure 1

PROCEDURE

1. Construct the circuit shown in Figure 1. Do not plug in the transformer at this time.

2. Have someone carefully check your wiring.

3. Connect Channel 1 of your oscilloscope across the lamp and Channel 2 across the SCR as shown in Figure 1. Note that the ground clip for each channel probe must be clipped between the lamp and SCR as shown. This will serve as a common ground connection and is necessary to avoid shorting the circuit between separate ground points. As a result of this common ground connection, the SCR voltage waveform will be inverted on the screen. Some

oscilloscopes have an invert button that can be used to correct this. Ask your instructor for assistance. In any case, the inversion of the SCR voltage waveform will not detract from you observing the relationship between the SCR and load voltage waveforms.

4. Plug in the transformer's power cord.

5. Adjust the variable resistor over its entire range and observe the lamp and SCR voltage waveforms on the oscilloscope.

   Note: The 470 Ω resistor can be replaced with a greater or lesser value to vary the control capability of the variable resistor.

6. Note which waveform represents the conduction angle and which represents the delay angle.

7. Note the relationship between conduction angle and lamp brightness.

8. Note the relationship between resistance of the variable resistor and conduction angle.

9. Determine the relationship between gate current and conduction angle.

10. Set the variable resistor so the conduction angle of each DC pulse is slightly greater than 90°.

11. Draw and label the two voltage waveforms on Screen 1 on the answer sheet. Clearly mark conduction angle and delay angle.

12. Remove diode D_2 and observe the voltage waveforms and lamp brightness.

13. Think of a way to obtain two ranges of brightness for the lamp.

14. Unplug the transformer's power cord and answer all remaining questions on the answer sheet.

NAME:________________________________CLASS:__________DATE:__________

QUESTIONS

1. What is the relationship between conduction angle and lamp brightness?
___

2. What is the minimum conduction angle? ___________

3. What is the maximum conduction angle? ___________

4. When the conduction angle is minimum, is the gate current (a) minimum, or (b) maximum? __
Why?
___

5. When the conduction angle is maximum, is the variable resistor set for (a) a greater resistance, or (b) a lesser resistance? __ Explain.
___

6. When D_2 was removed, how did this affect lamp brightness?
___

7. How can a switch be used to obtain two ranges of brightness for the lamp?
___

8. Explain how the average DC current in the lamp is related to the conduction angle of the SCR.
___
___

9. How does the SCR turn off once it begins to conduct?
___
___
___
___

10. Why is some gate current necessary to trigger the SCR?
___
___
___
___
___

Screen 1

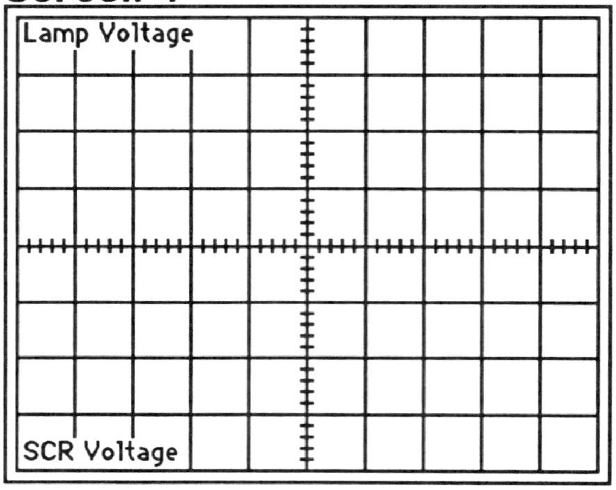

A TRIAC Circuit

EXPERIMENT
39

PURPOSE

The purpose of this experiment is to familiarize you with the operation of the gate-controlled thyristor called the TRIAC. Here, you will build and test a TRIAC control circuit.

FORMAL OBJECTIVES

After completing this experiment, you will be able to:

- recognize a TRIAC control circuit.
- use the oscilloscope to verify the operation of the TRIAC circuit.
- explain the operation of the TRIAC in terms of delay angle and conduction angle.

PREPARATIONS

In preparation for this experiment, you should study Section 16-4 of *Exploring Electronic Devices*. Also, you will need to solder wire leads on the terminals of your TRIAC so it can be plugged into your protoboard. Optionally, you can use clip leads.

COMPONENTS and EQUIPMENT

1 - Dual-trace oscilloscope with leads
1 - 25.2 V CT, 450 mA Transformer
1 - 2N6342/3 TRIAC, 200/400 V, 8 A
1 - 5 kΩ Trimmer potentiometer
1 - plug-in experiment board (breadboard)
1 - #1815 Lamp, 14 V, 0.2 A
1 - 100 Ω, 2 W Resistor
1 - 150 Ω, 1/4 W Resistor

DISCUSSION

In the following steps, you will assemble and test the TRIAC circuit of Figure 1. As you can see, 25.2 V RMS is applied to the TRIAC circuit from the secondary of the transformer. Therefore, the peak voltage across the TRIAC when it is not conducting is 25.2 V • 1.414 = 35.6 V_P, well below the V_{DRM} rating of the TRIAC.

Recall from your reading that the TRIAC will remain off until it is triggered on. The main-terminal voltage (V_{MT}) at which the TRIAC triggers is determined by the amount of gate current (I_G). As you can see in Figure 1, the gate current can be adjusted with the 5-kΩ potentiometer, wired as a rheostat. If the gate current is insufficient, the TRIAC will not trigger on. The potentiometer is used to vary the delay angle and angle of conduction for the TRIAC. In this experiment, you will note that, when the TRIAC is conducting, its conduction angle can only be varied over a range of 90° to 180° and 270° to 360°.

The TRIAC used in this experiment is either the 2N6342 (V_{DRM} = 200 V) or the 2N6343 (V_{DRM} = 400 V), or an equivalent replacement such as the Radio Shack part 276-1000. The maximum RMS on-state current ($I_{T(RMS)}$) for these devices is 8 A. The gate trigger current (I_{GT}) is in the neighborhood of 10 mA. The holding current (I_H) is typically 6 mA.

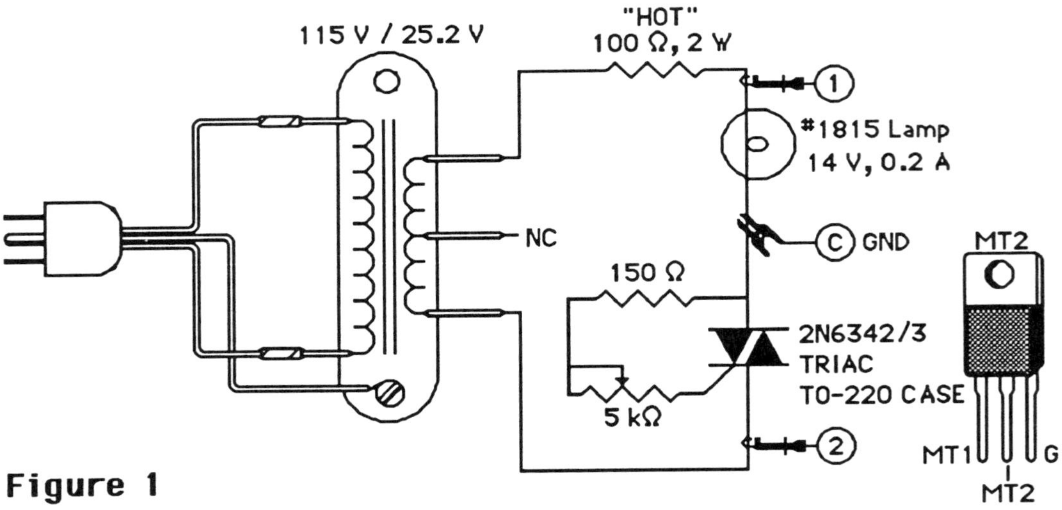

Figure 1

PROCEDURE

1. Construct the circuit shown in Figure 1. Do not plug in the transformer at this time.

2. Have someone carefully check your wiring.

3. Connect Channel 1 of your oscilloscope across the lamp and Channel 2 across the TRIAC as shown in Figure 1. Note that the ground clip for each channel probe must be clipped between the lamp and TRIAC as shown. This will serve as a common ground connection and is necessary to avoid shorting the circuit between separate ground points. As a result of this common ground connection, the TRIAC voltage waveform will be inverted on the screen.

Some oscilloscopes have an invert button that can be used to correct this. Ask your instructor for assistance. In any case, the inversion of the TRIAC voltage waveform will not detract from you observing the relationship between the TRIAC and load voltage waveforms.

4. Plug in the transformer's power cord.
5. Adjust the variable resistor over its entire range and observe the lamp and TRIAC voltage waveforms on the oscilloscope.
6. Note which waveform represents the conduction angle and which represents the delay angle.
7. Note the relationship between conduction angle and lamp brightness.
8. Note the relationship between resistance of the variable resistor and conduction angle.
9. Determine the relationship between gate current and conduction angle.
10. Set the variable resistor so the conduction angle of the positive alternation is slightly greater than 90° and the negative alternation is slightly greater than 270°.
11. Draw and label the two voltage waveforms on Screen 1 on the answer sheet. Clearly mark conduction angle and delay angle.
12. Unplug the transformer's power cord and answer all remaining questions on the answer sheet.

NAME:______________________ CLASS:__________ DATE:__________

QUESTIONS

1. What is the relationship between conduction angle and lamp brightness?

____

2. What are the minimum conduction angles? __________ and __________

3. What are the maximum conduction angles? __________ and __________

4. When the conduction angle is minimum, is the gate current (a) minimum, or (b) maximum? __
Why?

____

5. When the conduction angle is maximum, is the variable resistor set for (a) a greater resistance, or (b) a lesser resistance? __ Explain.

____

6. Explain how the average DC current in the lamp is related to the conduction angle of the TRIAC.

____
____
____
____

Screen 1

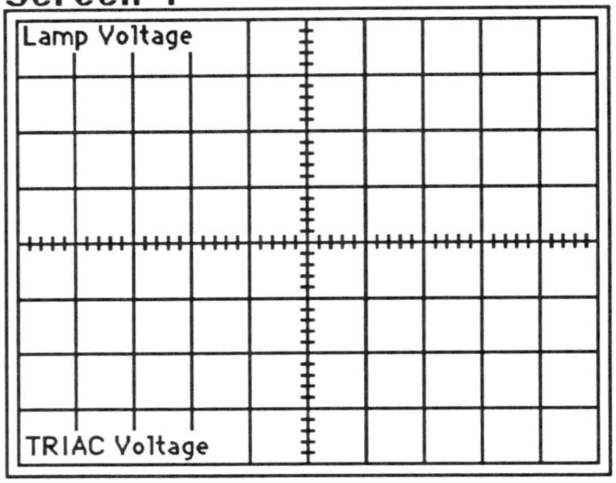

7. How does the TRIAC turn off once it begins to conduct?

____
____
____
____

8. Why is some gate current necessary to trigger the TRIAC?

____
____

9. In this experiment, the voltage across the TRIAC reaches a peak of about 36 V at minimum conduction angle (90°). Assume the total gate resistance to be the sum of the 150 Ω resistor and the resistance setting of the variable resistor. Determine the approximate peak gate current when the conduction angle is 90°.
$I_{G(90°)}$ = ____

The Unijunction Transistor

EXPERIMENT **40**

PURPOSE

The purpose of this experiment is to familiarize you with the operation of the unijunction transistor (UJT).

FORMAL OBJECTIVES

After completing this experiment, you will be able to:

- construct relaxation oscillator circuits using the unijunction transistor.
- analyze the characteristics of a UJT using an oscilloscope.
- explain the characteristics of the UJT.
- explain the operation of the relaxation oscillator.

PREPARATIONS

In preparation for this experiment, you should study Section 16-5 of *Exploring Electronic Devices*.

COMPONENTS and EQUIPMENT

1 - DC Supply (+12 V)
1 - Dual-trace oscilloscope with leads
1 - 0.1 μF Capacitor
1 - 100 Ω, 1/4 W Resistor
1 - plug-in experiment board (breadboard)
1 - 2N4851 UJT
1 - 20 kΩ, 1/4 W Resistor
1 - 20 Ω, 1/4 W Resistor

DISCUSSION

In the following steps, you will assemble and test the UJT relaxation oscillator shown in Figure 1. You should recall that the capacitor charges until V_P is reached. This peak voltage is the trigger point at which the emitter (E) to base 1 (B1) junction conducts heavily to quickly discharge the capacitor down to the valley voltage (V_V). At the valley voltage, the emitter current drops below the valley current level I_V, usually near 10 mA, and the E to B1 junction stops conducting. The capacitor once again begins to charge through resistor R, which in this case is 20 kΩ.

The peak voltage trigger point is determined largely by the intrinsic standoff ratio where: $\eta = r_{B1}/(r_{B1} + r_{B2})$.

The peak voltage must slightly exceed the sum of the E to B1 junction voltage (~0.7 V) and the voltage drop across the bulk resistance of base 1 (r_{B1}). The voltage across the B1 bulk resistance is equal to $V_S \bullet \eta = V_S \bullet r_{B1}/(r_{B1} + r_{B2})$. Thus, $V_P \sim 0.7\ V + (V_S \bullet r_{B1}/(r_{B1} + r_{B2}))$.

In this experiment, you will be able to measure the trigger voltage V_P and use this to determine the intrinsic standoff ratio of your UJT where: $V_{rB1} \sim V_P - 0.7\ V$ and

$$\eta = r_{B1}/(r_{B1} + r_{B2}) = V_{rB1}/V_S = (V_P - 0.7\ V) / V_S$$

Since V_S is known, the intrinsic standoff ratio can be determined after V_P is measured. As mentioned is the text, we are assuming the E to B1 junction voltage to be 0.7 V when V_P is reached. In reality, the E to B1 junction voltage may be less than this amount depending on device and temperature.

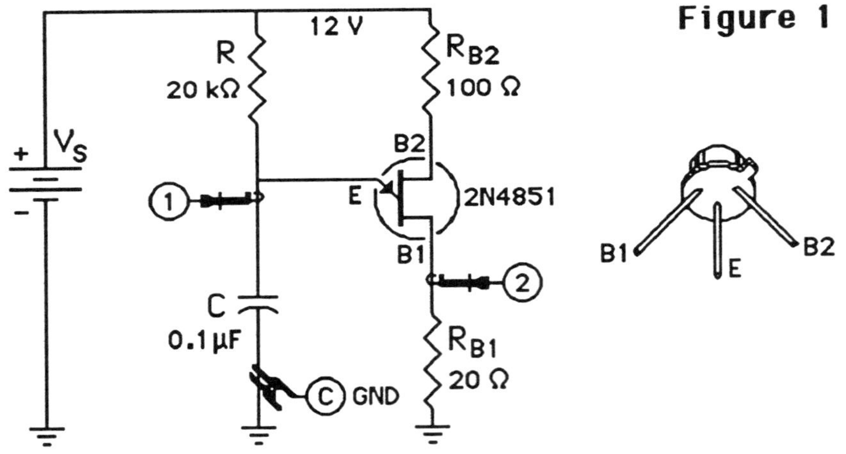

Figure 1

PROCEDURE

1. Construct the circuit shown in Figure 1. Check your wiring. Make sure B1 is connected to the 20 Ω resistor and E to the capacitor.

2. Connect Channel 1 of your oscilloscope across the capacitor and Channel 2 across the 20 Ω resistor.

3. Apply +12 V to your circuit and adjust the oscilloscope controls to view the capacitor voltage waveform in the upper half of the screen and the base 1 waveform in the lower half of the screen. You should be able to view two to four cycles across the screen.

4. Momentarily set the input mode-select switches for both channels to the GND position to establish a 0 V DC reference line on the screen. Use the center line for the upper half of the screen (CH-1) and the bottom line for the lower half of the screen (CH-2). Now set the mode-select switches to the DC position.

5. With the input mode-select switches in the DC position, carefully measure the peak and valley voltages of the capacitor voltage waveform. These will be positive voltages. Record your findings on the answer sheet.

6. Using the measured value of peak voltage, calculate the intrinsic standoff ratio (η). Show your work on the answer sheet.

7. Measure and record the peak-pulse voltage at B1 (CH-2).

8. Use the peak-pulse voltage measured at B1 to calculated the peak E to B1 current (I_{EB1}). This is the capacitor discharge current which forces the UJT deep into saturation for a very short time. Very quickly, I_{EB1} drops below the valley current point (I_V) and the UJT turns off (resets). You will see that the peak discharge current is very much greater than the valley current.

9. Draw the capacitor and B1 voltage waveforms on Screen 1 on the answer sheet. Label the peak and valley voltage points and indicate voltages.

10. Remove power and answer all remaining questions.

NAME:______________________________CLASS:__________DATE:__________

QUESTIONS

1. V_P = __________V, V_V = __________V

2. $\eta = V_{rB1}/V_S = (V_P - 0.7\text{ V}) / V_S =$ ____________________________

3. V_{B1} = __________V

4. Peak E to B1 current = ____________________________

5. What are the factors that affect the frequency of oscillation in the UJT relaxation oscillator?
   1. ____________________________
   2. ____________________________
   3. ____________________________

6. How can the frequency of oscillation be changed?

7. Define the peak voltage point (V_P). ____________________________

8. What does the intrinsic standoff ratio of the UJT have to do with the peak voltage point?

9. Is the capacitor voltage waveform a (a) triangle wave, (b) square wave, or (c) a sawtooth wave? ___

10. How does the peak E to B1 current during capacitor discharge compare to the valley point current?

Screen 1

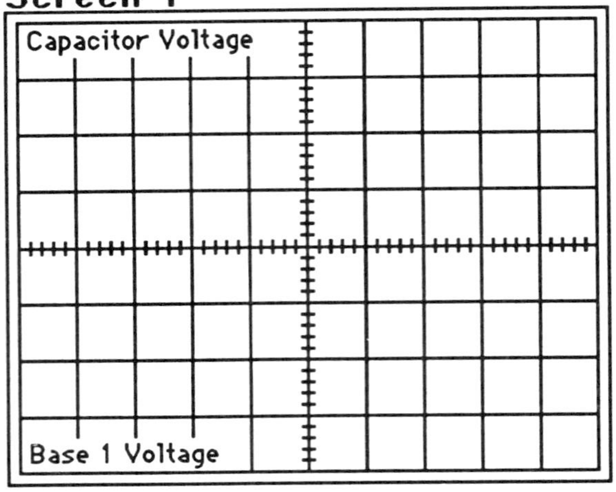

Linear Voltage Regulator

EXPERIMENT 41

PURPOSE

The purpose of this experiment is to familiarize you with the operation of a basic discrete-component linear voltage regulator.

FORMAL OBJECTIVES

After completing this experiment, you will be able to:

- construct a basic discrete-component linear voltage regulator.
- analyze a basic discrete-component linear voltage regulator using a multimeter.
- explain the voltage regulating and current limiting operation of a basic regulator.
- explain the limitations of a basic discrete-component linear voltage regulator.

PREPARATIONS

In preparation for this experiment, you should study Sections 17-1 and 17-2 of *Exploring Electronic Devices*.

COMPONENTS and EQUIPMENT

1 - Analog or digital multimeter
1 - Transformer, 115 V / 25.2 V CT
1 - 1,000 μF, 25 V Capacitor
1 - 2N2219 NPN Transistor
1 - 0.1 μF Film capacitor
1 - 100 Ω, 2 W Resistor
1 - 10 Ω, 1/4 W, 2% Resistor
1 - Protoboard
2 - 1N4001 Diodes
1 - 1N4742, 12 V, 1 W Zener diode
1 - 2N3904 NPN Transistor
1 - 220 Ω, 2 W Resistor
1 - 270 Ω, 1/4 W Resistor

DISCUSSION

In the following steps, you will assemble and test the basic discrete-component linear voltage regulator shown in Figure 1. As you can see, this is a basic zener-controlled series-pass regulator with current limiting capability. Q_1 is the series-pass active device use for voltage control and Q_2 is the current-limit transistor. When the load is such that the voltage drop across the 10-Ω limit resistor is approximately 0.7 V, Q_2 begins to conduct and reduces the drive to Q_1.

The zener voltage reference is 12 V. Recall there is an approximate 0.7-V drop across the BE junction of Q_1. Thus the regulated output voltage is 12 V - 0.7 V - V_{Rlimit}. You will discover that the no-load to full-load regulation of this circuit is not all that good because of the voltage drop across the limit resistor and slight variations in V_{BE} of Q_1. In the next experiment, we will correct this problem using voltage feedback. Therefore, when you finish this experiment, do not disassemble your circuit. Simply modify it for the next experiment.

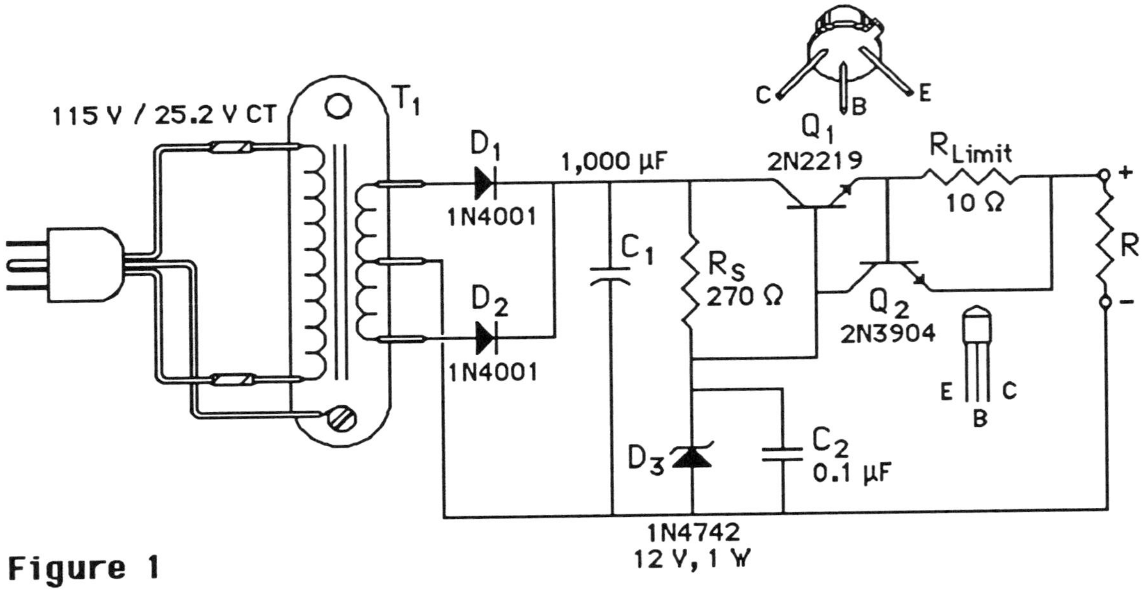

Figure 1

PROCEDURE

1. Before doing anything else, take time to mathematically analyze the circuit of Figure 1. Show all work and answers on the answer sheet. The secondary voltage is 25.2 V RMS total. Refer to Sections 17-1 and 17-2 for assistance. Using reference material is a necessary part of being a technician.

2. Construct the circuit shown in Figure 1. *Do not* even think of plugging in the transformer *until* another person has thoroughly checked your wiring.

   Do not place a load resistor across the output at this time.

3. Plug in the transformer's power cord and use your DC voltmeter to measure V_{C1}, $V_{Z(D3)}$, and V_{out}. Record your readings on the answer sheet.

4. Connect the 220 Ω, 2 W resistor across the output of your regulator and measure V_{C1}, $V_{Z(D3)}$, and V_{out} again. Record your readings on the answer sheet.

5. Measure the voltage drop across R_{limit} and the voltage across Q_1. Record your readings on the answer sheet.

6. Use the data collected above to calculate I_{load}, P_{load}, and P_{Q1}. Show your work on the answer sheet.

7. Use the output voltage measured in Step 3 (V_{NL}) and the output voltage measured in Step 4 (V_{FL}) to determine the percent regulation. Show your work on the answer sheet.

8. Use P_{load} and P_{Q1} to calculate the percent efficiency of this regulator. Show your work on the answer sheet.

9. Remove the 220 Ω load resistor and replace it with the 100 Ω, 2 W resistor. Measure V_{C1}, $V_{Z(D3)}$, V_{out}, V_{Rlimit}, and $V_{CE(Q1)}$. Record your readings on the answer sheet.

10. Unplug the transformer's power cord and answer all remaining questions on the answer sheet.

DO NOT DISASSEMBLE YOUR CIRCUIT.
YOU WILL MODIFY IT FOR THE NEXT EXPERIMENT.

NAME:______________________________CLASS:__________DATE:__________

QUESTIONS

Calculated Values:

1. V_{C1} = ____________________
2. V_{out} < ____________________
3. I_{Limit} = ____________________
4. $V_{CE(Q1)}$ = ____________________

Measured Values:

5. V_{C1} = __________V, $V_{Z(D3)}$ = __________V, V_{out} = __________V
6. With the 220 Ω load resistor: V_{C1} = __________V, $V_{Z(D3)}$ = __________V
   V_{out} = __________V
7. V_{Rlimit} = __________V, $V_{CE(Q1)}$ = __________V

Calculated Values:

8. I_{load} = ____________________
   P_{load} = ____________________
   P_{Q1} = ____________________
9. % Regulation = 100% • $(V_{NL} - V_{FL}) / V_{FL}$ = ____________________
   Note: V_{NL} = no-load output voltage and V_{FL} = full-load output voltage.
10. % Efficiency = 100% • $P_{load} / (P_{load} + P_{Q1})$ = ____________________

Measured Values with 100 Ω load resistor:

11. V_{C1} = __________V, $V_{Z(D3)}$ = __________V, V_{out} = __________V
    V_{Rlimit} = __________V, $V_{CE(Q1)}$ = __________V
12. Why is the output voltage with the 100 Ω load resistor so low?
    ____
    ____
13. Calculate I_{load} with the 100 Ω load resistor. I_{load} = ____________________
14. What is the maximum amount of output current this regulator will supply? Demonstrate this mathematically using the measured V_{Rlimit}. ____________________
15. What factors contributed to the percent regulation being so high in Question 9?
    ____
    ____

DO NOT DISASSEMBLE YOUR CIRCUIT. YOU WILL MODIFY IT FOR THE NEXT EXPERIMENT.

Linear Voltage Regulator with Voltage Feedback

EXPERIMENT

42

PURPOSE

The purpose of this experiment is to familiarize you with the use of voltage feedback in a basic discrete-component linear voltage regulator.

FORMAL OBJECTIVES

After completing this experiment, you will be able to:

• construct a basic discrete-component linear voltage regulator employing voltage feedback.
• analyze a basic discrete-component linear voltage regulator using a multimeter.
• explain the voltage regulating and current limiting operation of a basic regulator that employs voltage feedback.

PREPARATIONS

In preparation for this experiment, you should study Section 17-2 of *Exploring Electronic Devices*.

COMPONENTS and EQUIPMENT

1 - Analog or digital multimeter
1 - Transformer, 115 V / 25.2 V CT
1 - 1,000 μF, 25 V Capacitor
1 - 2N2219 NPN Transistor
1 - LM741 Op-Amp
1 - 0.1 μF Film capacitor
1 - 100 Ω, 2 W Resistor
1 - 10 Ω, 1/4 W, 2% Resistor
1 - Protoboard
2 - 1N4001 Diodes
1 - 1N4742, 12 V, 1 W Zener diode
1 - 2N3904 NPN Transistor
1 - 1 kΩ, 1/4 W Resistor
1 - 220 Ω, 2 W Resistor
1 - 270 Ω, 1/4 W Resistor

DISCUSSION

In the following steps, you will modify your regulator from the previous experiment by adding resistor R_B and the 741 op-amp as shown in Figure 1. Then, you will perform the same measurements and tests as before to compare the performance of this improved regulator with the old circuit. You will see a significant improvement in % regulation.

The op-amp in Figure 1 acts as an extremely high-gain amplifier that is used to compare the actual load voltage with the zener reference voltage. If the output voltage was too low, the zener voltage seen at the noninverting input (pin 3) would be greater than the feedback voltage at the inverting input (pin 2). Even a slight difference between these two voltages would be greatly amplified by the op-amp producing a very high positive voltage at pin 6 to drive R_B and Q_1. This high drive voltage, and current, causes Q_1 to conduct more heavily which increases the output voltage. However, the output voltage will never be greater than the zener reference voltage. The very high gain of the op-amp guarantees that the output voltage will not drop down, under any load condition, unless the load is equal to or greater than the current limit. Note that the current limit transistor Q_2 will still function to limit the drive to the base of Q_1 if the load is equal to or greater than the current limit as determined by the limit resistor R_{Limit}. Thus, this regulator has excellent regulation and current limiting characteristics.

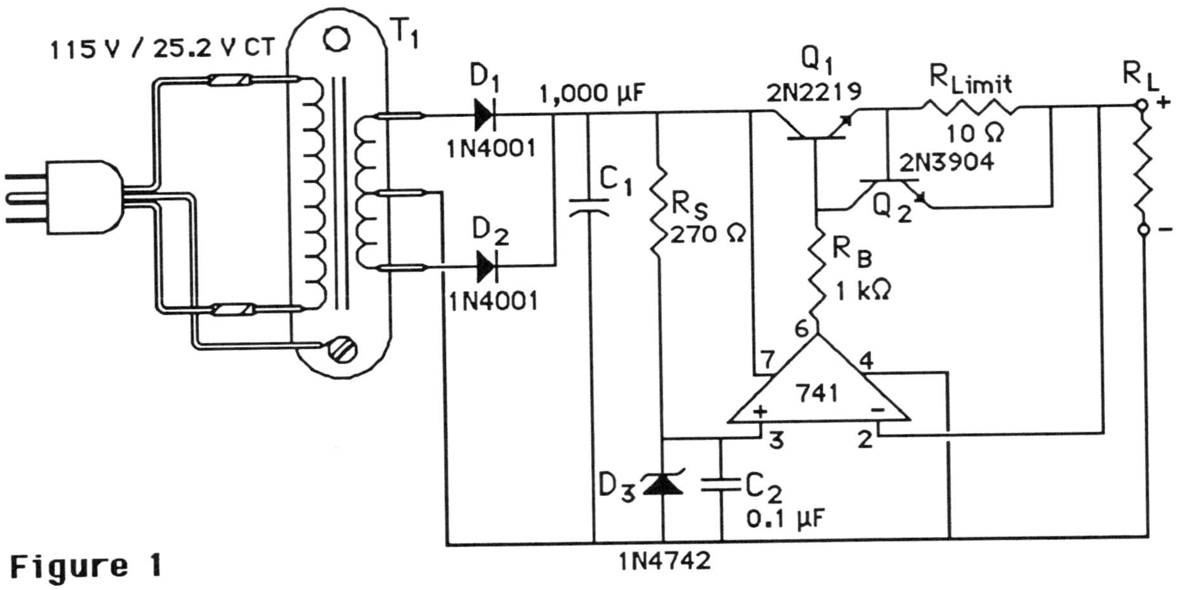

Figure 1

PROCEDURE

1. Construct the circuit shown in Figure 1. *Do not* even think of plugging in the transformer *until* another person has thoroughly checked your wiring.

   Do not place a load resistor across the output at this time.

2. Plug in the transformer's power cord and use your DC voltmeter to measure V_{C1}, $V_{Z(D3)}$, V_{out}, and $V_{PIN\ 6}$. Record your readings on the answer sheet.

3. Connect the 220 Ω, 2 W resistor across the output of your regulator and measure V_{C1}, $V_{Z(D3)}$, V_{out}, and $V_{PIN\ 6}$ again. Record your readings on the answer sheet.

4. Measure the voltage drop across R_{limit} and the voltage across Q_1. Record your readings on the answer sheet.

5. Use the data collected above to calculate I_{load}, P_{load}, and P_{Q1}. Show your work on the answer sheet.

6. Use the output voltage measured in Step 2 (V_{NL}) and the output voltage measured in Step 3 (V_{FL}) to determine the percent regulation. Show your work on the answer sheet.

7. Use P_{load} and P_{Q1} to calculate the percent efficiency of this regulator. Show your work on the answer sheet.

8. Remove the 220 Ω load resistor and replace it with the 100 Ω, 2 W resistor. Measure V_{C1}, $V_{Z(D3)}$, V_{out}, V_{Rlimit}, $V_{CE(Q1)}$, and $V_{PIN\ 6}$. Record your readings on the answer sheet.

9. Set your multimeter to measure current in the 100 mA or 200 mA range. Place this meter in series with the 100 Ω load resistor to measure the load current. Record your reading on the answer sheet.

10. Remove the 100 Ω load resistor and place a dead short across the output terminals. Measure V_{C1}, $V_{Z(D3)}$, V_{out}, V_{Rlimit}, $V_{CE(Q1)}$, and $V_{PIN\ 6}$. Record your readings on the answer sheet.

11. Unplug the transformer's power cord and answer all remaining questions on the answer sheet.

NAME:______________________________ CLASS:__________ DATE:__________

QUESTIONS

Measured Values:

1. V_{C1} = __________V, $V_{Z(D3)}$ = __________V, V_{out} = __________V
   $V_{PIN\ 6}$ = __________V
2. With the 220 Ω load resistor: V_{C1} = __________V, $V_{Z(D3)}$ = __________V
   V_{out} = __________V, $V_{PIN\ 6}$ = __________V
3. V_{Rlimit} = __________V, $V_{CE(Q1)}$ = __________V

Calculated Values:

4. I_{load} = ______________________________
   P_{load} = ______________________________
   P_{Q1} = ______________________________
5. % Regulation = 100% • $(V_{NL} - V_{FL}) / V_{FL}$ = ______________________________
   Note: V_{NL} = no-load output voltage and V_{FL} = full-load output voltage.
6. % Efficiency = 100% • $P_{load} / (P_{load} + P_{Q1})$ = ______________________________

Measured Values with 100 Ω load resistor:

7. V_{C1} = __________V, $V_{Z(D3)}$ = __________V, V_{out} = __________V
   V_{Rlimit} = __________V, $V_{CE(Q1)}$ = __________V, $V_{PIN\ 6}$ = __________V
8. I_{Load} = __________mA

9. Measured Values (with a dead short at the output terminals):
   V_{C1} = __________V, $V_{Z(D3)}$ = __________V, V_{out} = __________V
   V_{Rlimit} = __________V, $V_{CE(Q1)}$ = __________V, $V_{PIN\ 6}$ = __________V

10. Compare the data collected above with the data of the previous experiment:
    How do the no-load output voltages compare? ______________________________
    How do the full-load output voltages compare? ______________________________
    Which regulator circuit has the lowest percent regulation? ______________________________
    Which is the most efficient regulator? ______________________________
    Which regulator is least affected by changes in load? ______________________________
    Which regulator circuit is the best? ______________________________
11. What happens to $V_{PIN\ 6}$ as the load is increased? ______________________________ Why?
    ____
    ____

12. Explain why the regulator circuit of this experiment is the best. Include a discussion of voltage feedback.
    ____
    ____
    ____
    ____

The 723 Voltage Regulator

EXPERIMENT

43

PURPOSE

The purpose of this experiment is to familiarize you with the LM723 linear voltage regulator. In this experiment, you will build a voltage regulator that can be set for any output voltage between 7.15 V and +Vin - 3 V. The output voltage selected for this circuit is +12 V.

FORMAL OBJECTIVES

After completing this experiment, you will be able to:

- construct a voltage regulator using the LM723.
- analyze an LM723 regulator circuit using a multimeter.
- explain the voltage regulating and current limiting operation of an LM723 regulator circuit.

PREPARATIONS

In preparation for this experiment, you should study Section 17-3 of *Exploring Electronic Devices*.

COMPONENTS and EQUIPMENT

1 - Analog or digital multimeter
1 - Transformer, 115 V / 25.2 V CT
1 - 1,000 μF, 25 V Capacitor
1 - 100 pF Capacitor
1 - 6.8 kΩ, 1/4 W Resistor
1 - 220 Ω, 2 W Resistor
1 - Protoboard
2 - 1N4001 Diodes
1 - LM723 Regulator IC
1 - 10 kΩ, 1/4 W Resistor
1 - 10 Ω, 1/4 W, 2% Resistor
1 - 100 Ω, 2 W Resistor

DISCUSSION

In the following steps, you will build and test the voltage regulator circuit of Figure 1. In this circuit, the LM723 linear regulator IC takes the place of most of the regulator circuitry of the previous experiment. The zener voltage reference, high-gain op-amp, current-limit transistor, and series-pass transistor are all inside the 723 IC. Only a few additional components are needed to establish the current limit (R_{Limit}) and the feedback voltage ratio (R_1 and R_2). The 100 pF capacitor is necessary to stabilize the internal op-amp and prevent it from oscillating.

The internal zener reference voltage, which can be measured at pin 6, is 7.15 V. The feedback voltage divider permits an output voltage anywhere in the range of +7.15 V to $+V_{IN}$ minus approximately 3 V for a minimum input-output voltage differential. If desired, R_1 can be a 10 kΩ potentiometer wired as a rheostat to allow variable voltage adjustment over the range of +7.15 V to +14.3 V. The 723 itself can supply a maximum of 150 mA. If greater current is needed, an external power transistor can be used and R_{Limit} selected accordingly. See Section 17-3 of *Exploring Electronic Devices*.

Figure 1

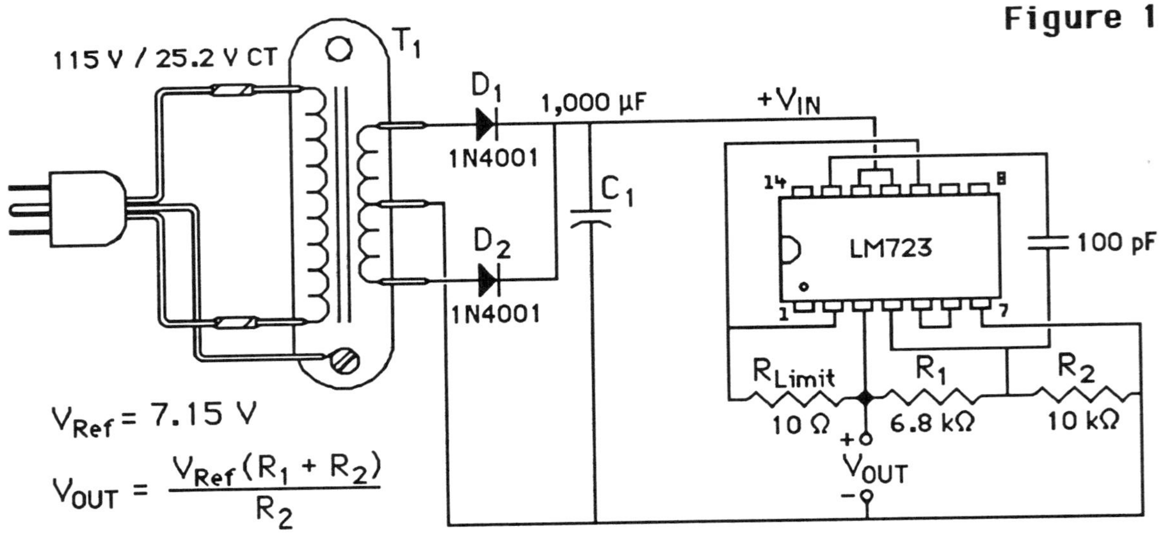

$V_{Ref} = 7.15$ V

$$V_{OUT} = \frac{V_{Ref}(R_1 + R_2)}{R_2}$$

PROCEDURE

1. Construct the circuit shown in Figure 1. *Do not* even think of plugging in the transformer *until* another person has thoroughly checked your wiring.

   Do not place a load resistor across the output at this time.

2. Plug in the transformer's power cord and use your DC voltmeter to measure V_{C1}, $V_{PIN\ 4}$, $V_{PIN\ 6}$ and V_{OUT}. Record your readings on the answer sheet.

3. Connect the 220 Ω, 2 W resistor across the output of your regulator and measure V_{C1} and V_{OUT} again. Record your readings on the answer sheet.

4. Measure the voltage drop across R_{limit}. Record your readings on the answer sheet.

5. Use the data collected above to calculate I_{load}, P_{load}, and P_{LM723}. Show your work on the answer sheet.

6. Use the output voltage measured in Step 2 (V_{NL}) and the output voltage measured in Step 3 (V_{FL}) to determine the percent regulation. Show your work on the answer sheet.

7. Use P_{load} and P_{LM723} to calculate the percent efficiency of this regulator. Show your work on the answer sheet.

8. Remove the 220 Ω load resistor and replace it with the 100 Ω, 2 W resistor. Measure V_{C1}, V_{Rlimit}, and V_{OUT}. Record your readings on the answer sheet.

9. Set your multimeter to measure current in the 100 mA or 200 mA range. Place this meter in series with the 100 Ω load resistor to measure the load current. Record your reading on the answer sheet.

10. Unplug the transformer's power cord and answer all remaining questions on the answer sheet.

NAME:______________________CLASS:__________DATE:__________

QUESTIONS

Measured Values:

1. V_{C1} = __________V, $V_{PIN\,4}$ = __________V, $V_{PIN\,6}$ = __________V
   V_{OUT} = __________V
2. Why is $V_{PIN\,4} = V_{PIN\,6}$? ______________________________
   ____
3. With the 220 Ω load resistor: V_{C1} = __________V, V_{OUT} = __________V
4. V_{Rlimit} = __________V

Calculated Values:

5. I_{load} = ______________________________
   P_{load} = ______________________________
   P_{LM723} = ______________________________
6. % Regulation = 100% • $(V_{NL} - V_{FL}) / V_{FL}$ = ______________________________
   Note: V_{NL} = no-load output voltage and V_{FL} = full-load output voltage with the 220 Ω R_L.
7. % Efficiency = 100% • $P_{load} / (P_{load} + P_{LM723})$ = ______________________________

8. Measured Values with 100 Ω load resistor:
   V_{C1} = __________V, V_{OUT} = __________V, V_{Rlimit} = __________V
   I_{Load} = __________mA

9. Compare the data collected above with the data of the previous experiment:
   How do the no-load output voltages compare? ______________________________
   How do the full-load output voltages compare? ______________________________
   Which regulator circuit has the lowest percent regulation? ______________________________
   Which is the most efficient regulator? ______________________________
   Which regulator is least affected by changes in load? ______________________________
   Which regulator circuit is the best? ______________________________

10. What value should R_1 be if the desired output voltage is +9 V? Assume R_2 is 10 kΩ. Show your work.
    ____

11. What value should R_{Limit} be if the current is to be limited to 100 mA? Show your work.
    ____

An Optoisolator Circuit

EXPERIMENT
44

PURPOSE

The purpose of this experiment is to introduce you to optoisolators. The MOC3010 LATRIAC output optoisolator will be used in this experiment to accomplish that purpose.

FORMAL OBJECTIVES

After completing this experiment, you will be able to:

- construct basic optoisolator circuits using the MOC3010.
- explain the operation of the MOC3010 TRIAC optoisolator and optoisolators in general.

PREPARATIONS

In preparation for this experiment, you should study Section 18-4 of *Exploring Electronic Devices*.

COMPONENTS and EQUIPMENT

1 - Multimeter with leads
1 - Transformer, 115 V/25.2 V CT
1 - 2N6342/3 TRIAC
1 - 1N4001 Diode
1 - 150 Ω, 1/4 W Resistor
1 - 100 μF, 50 V Electrolytic capacitor
1 - Plug-in experiment board (breadboard)
1 - MOC3010 TRIAC Optoisolator
1 - #1815 Lamp, 14 V, 0.2 A
1 - 2 kΩ, 1 W Resistor
1 - 100 Ω, 2 W Resistor

DISCUSSION

In the following steps, you will assemble and test the optoisolator circuit shown in Figure 1. The MOC3010 optoisolator has an LED input and a LATRIAC output. The LATRIAC is used to trigger a larger power TRIAC. This circuit is not intended for brightness control. It is merely an on/off control circuit. The optoisolator is intended to isolated the high-voltage load circuit from a sensitive low-voltage control circuit - such as a 5 V computer logic control circuit. The circuit of Figure 1 is simply a demonstration circuit. The 25.2 V AC source is used for both the control side and the load side of the circuit - which would not be the case in a real application.

The jumper wire is used to activate and deactivate the circuit. When the jumper wire is in, the internal LED is activated. When the jumper is removed, the LED circuit is broken and the internal LATRIAC and external TRIAC will not trigger on. The manufacturer guarantees that the LATRIAC will trigger on with an LED current of no more than 15 mA. This is known as the LED trigger current (I_{FT}). The maximum forward current rating of the LED is 60 mA. Therefore, the operating forward current for the LED should be be in the range of 15 mA to 60 mA. The 2 kΩ resistor in Figure 1 insures that this is so.

The LATRIAC is rated at 250 V (V_{DRM}) and a power dissipation of 300 mW at 25°C. This device is not intended for direct load applications. It is intended only as a trigger for high-power TRIACs. The MOC3010 is part of a family of optoisolators that share the same general characteristics (MOC3009/10/11/12).

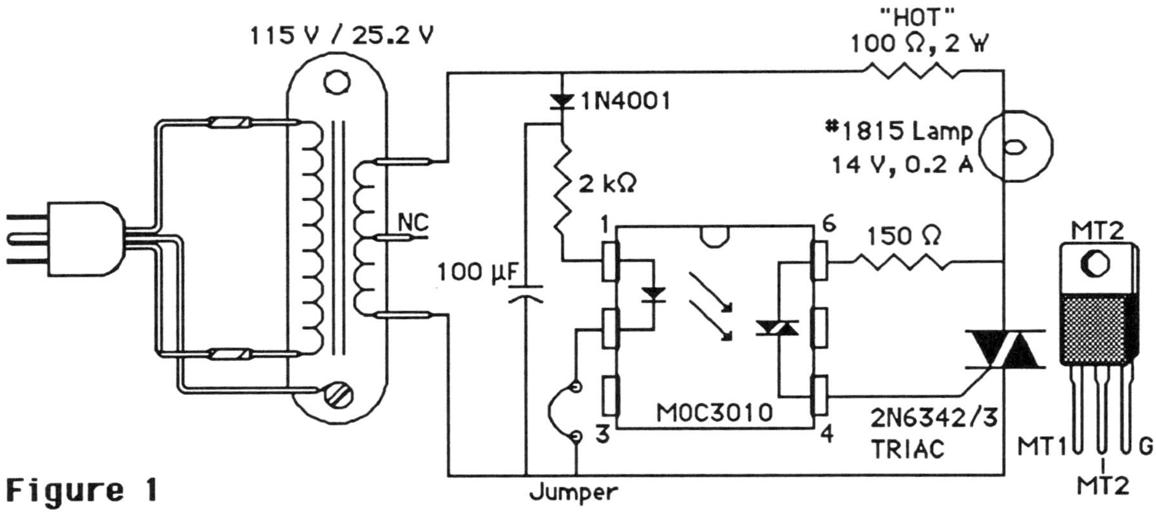

Figure 1

PROCEDURE

1. Construct the circuit shown in Figure 1. Do not plug in the transformer's power cord at this time. Do not insert the jumper at this time.

2. Have someone else check your wiring.

3. Plug in the transformer's power cord. The lamp should not come on.

4. Plug in the jumper wire. The lamp should now light.
5. Determine the forward current for the LED. Assume the LED has a forward voltage of 1.5 V.
6. Remove the jumper wire. The lamp should turn off.
7. Unplug the transformer's power cord and answer all remaining questions on the answer sheet.

NOTES AND CALCULATIONS

NAME:______________________________CLASS:__________DATE:_________

QUESTIONS

1. $I_F = (V_P - V_F) / 2\ k\Omega = ((25.2\ V_{RMS} \bullet 1.414) - 1.5\ V) / 2\ k\Omega =$ __________mA

2. After the jumper wire is removed, when does the lamp actually go off?

   ____

3. Name some applications for optoisolators such as the MOC3010.

   ____

   ____

3. Use the space below to redraw the optoisolator circuit using a separate +5 VDC control circuit. In other words, a +5 V source is used to energize the LED. Label the value of the current limit resistor for the LED. Assume the forward voltage for the LED is 1.5 V.

The Electronics Lab

APPENDIX A

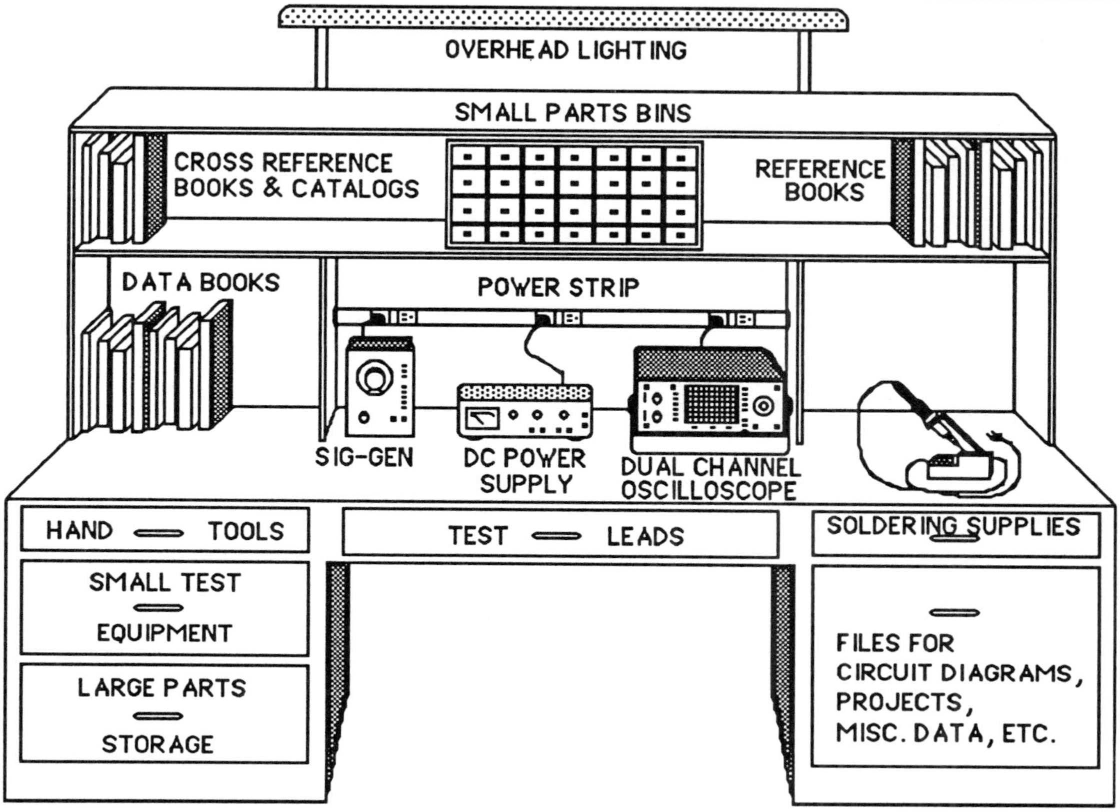

HAND TOOLS
needle nose pliers
diagonal wire cutters
set of small screw drivers
set of nut drivers
razor knife
wire stripper
forceps
set of Philips screw drivers
set of plastic tuning tools
misc. wire-wrap tools

SMALL TEST EQUIPMENT
analog multimeter
digital multimeter
capacitance meter
digital test probe
digital pulser probe
clamp-on ammeter

TEST LEADS
clip-lead oscilloscope cables
10:1 low-capacitance probe
red/black test leads
colored clip leads
multimeter test leads
protoboard (breadboard)

SOLDERING SUPPLIES
60/40 rosin-core solder
a tin of paste flux
solder wick for desoldering
solder sucker for desoldering
clip-on heat sink
cotton swabs and alcohol
soldering iron replacement tips
small fine-grade file
fine-grade sandpaper
misc. hookup wire

Shop Safety

APPENDIX B

Why do even well experienced technicians and engineers fall victum to death by electrocusion? The answer is simple: experience and seniority are no substitutes for a conscious application of safety precautions. Keep these basic precautions in mind when working on equipment:

1. **Be aware of all high-voltage points in a circuit or system.**
2. **Make sure all electrical equipment is properly grounded through a common earth ground.**
3. **Wear insulated shoes and clothing when working around high voltage.**
4. **Make high-voltage measurements with one hand keeping all other parts of the body clear of any ground points, such as metal chassis.**
5. **Do not make high-voltage measurements by yourself. Another person should be present to aid you if you run into trouble.**
6. **Continue to review in your mind exactly where high-voltage points in the circuit are and stay clear. Do not become careless. Plan your moves and be deliberate in your actions.**
7. **Do not work on, or around, a circuit that has lethal voltage applied unless it is absolutely necessary.**
8. **Turn off main circuit breakers and tag them with your signature while working on electrical equipment. You are the only person who should remove the tag and reset the circuit breaker.**
9. **Use a grounding stick to make sure all high-voltage capacitors are discharged before beginning any maintenance routine. A grounding stick is a metal shaft with a large insulated handle. The metal shaft is grounded to the equipment chassis via a flexible braided cable.**
10. **Do not work when you are overly tired. Be alert. Do not take shortcuts. Do not take chances.**

THINK SAFETY! PRACTICE SAFETY! LIVE LONG!

The Plug-in Experiment Board

APPENDIX
C

The plug-in breadboard (prototype board) is almost an absolute necessity. It allows you to quickly and neatly develop a circuit or perform an experiment with the minimum of frustration. You must understand how it is made before you can use it. Study the diagram below. As you can see, the top and bottom rows are electrical buses that can be used for common ground and ± voltage. The vertical columns of contacts are for component interconnection.

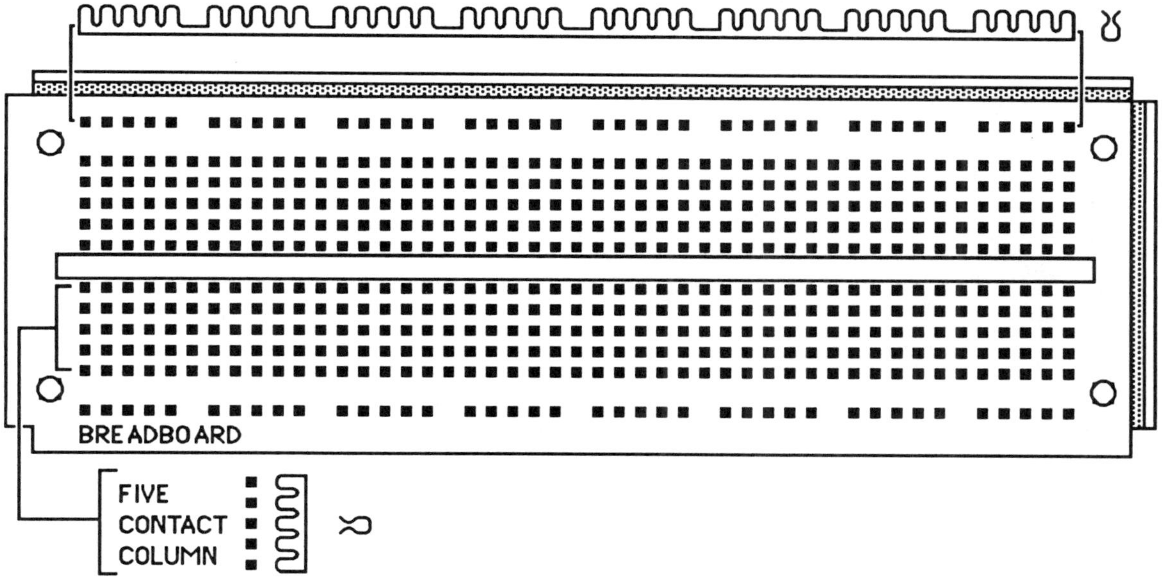

The DC Power Supply

APPENDIX

D

A good shop power supply will offer you at least the flexibility of the one shown here. Two variable DC voltage sources are available to be used at the same time or independently. The two output terminals for each variable DC source are each isolated outputs. Neither terminal is connected to ground. This permits you to place the two variable DC sources in parallel, for more current, or in series, for more voltage. The B supply can be made to track the A supply. When the slide switch is in the tracking position, the A voltage control is used to adjust the voltage of both supplies at the same time. This is very useful when paralleling the two supplies or for balanced bipolar supply requirements. Usually, supplies of this type can supply only about 500 mA or so which is sufficient for most experimentation and prototyping.

A third and separate DC supply is also available for use and is labeled as supply C, +5 V. This is a fixed voltage supply that can usually supply 5 A of current. This supply is generally used for digital type circuitry.A volt/ammeter is used to monitor current and voltage of any of the three DC sources. The selector switch makes voltage and current readings very convenient.

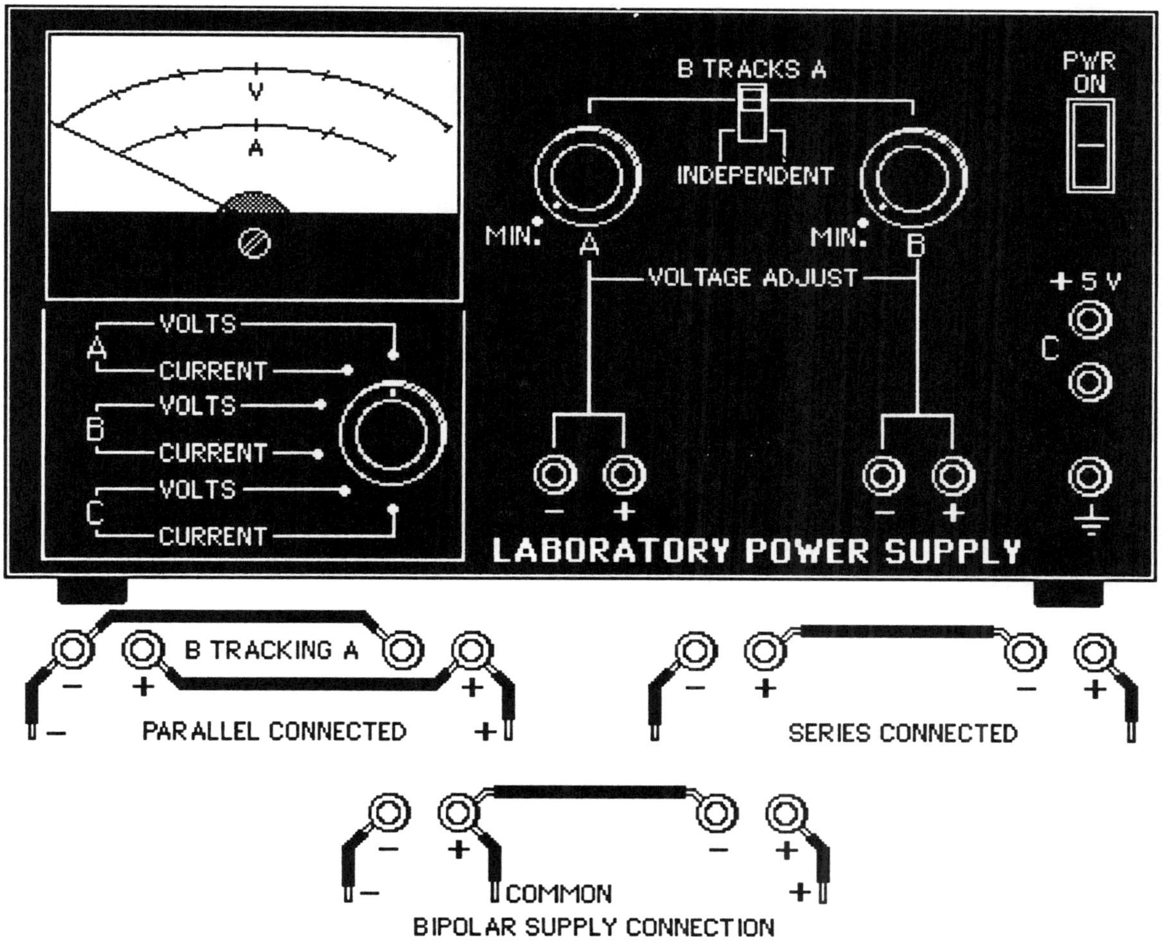

Using The Multimeter

APPENDIX E

The purpose of this appendix is to provide you with a summary of procedures to follow when using a multimeter. Unless specified, these procedures apply equally to analog and digital multimeters. For specific operating instructions and explanations of special functions, you will have to consult the operating manual for the multimeter you are using.

USING THE MULTIMETER TO MEASURE COMPONENT RESISTANCE

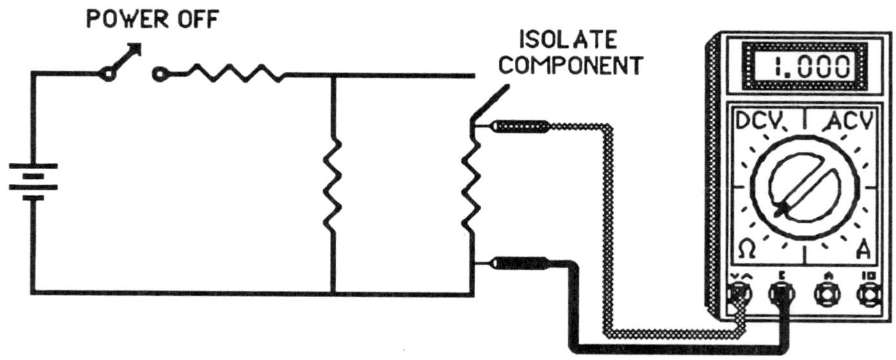

1. Remove circuit power.
2. Isolate the component to be measured by disconnecting at least one of its leads.
3. Select the Ω function. If the ohmmeter is analog, you will have to short the test leads together and use the ohms-zero-adjust knob to zero the pointer on the right end of the scale.
4. Connect the test leads across the component.
5. If the multimeter is not autoranging, start on the highest range and switch down to lower ranges until an in-range reading (digital) is obtained (mid- to right-scale reading for greatest accuracy when using an analog multimeter).

USING THE MULTIMETER TO MEASURE VOLTAGE

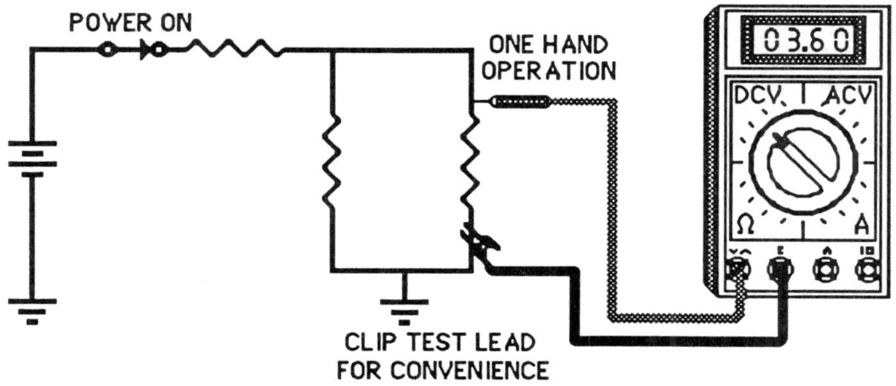

1. Think safety first! Be careful not to come in contact with lethal voltages. Find out where all high-voltage points are by examining circuit documentation.
2. Select the V function, AC or DC.
3. If the multimeter is not autoranging, start on the highest range and switch down to lower ranges until an in-range reading (digital) is obtained (mid- to right-scale reading for greatest accuracy when using an analog multimeter).
4. Apply power to the circuit.
5. Be careful to observe polarity when measuring DC with an analog multimeter to avoid damage to the meter (red lead to positive and black lead to negative). A digital multimeter will indicate polarity on the display. No harm will come to a digital multimeter if polarity is not observed.
6. Place the test leads in parallel with the component whose voltage is to be measured. A clip lead can be used for a common ground connection to the meter so measurements can be performed with one hand (for safety and convenience).

USING THE MULTIMETER TO MEASURE CURRENT

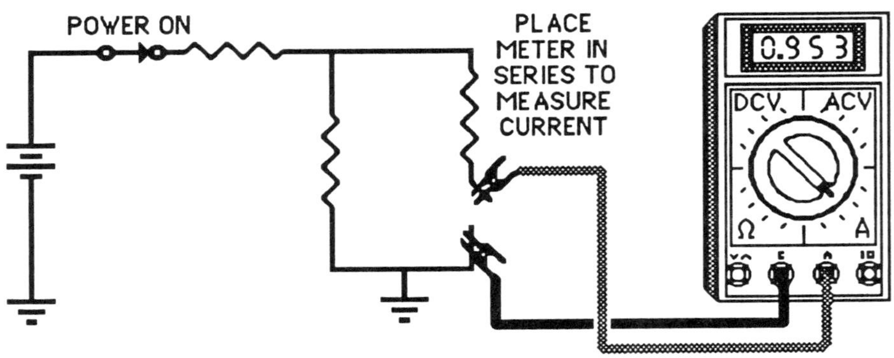

1. Think safety first! Be careful not to come in contact with lethal voltages. Find out where all high-voltage points are by examining circuit documentation.
2. Make sure circuit power is OFF before you begin.
3. Select the A function, AC or DC.
4. Open the circuit at the point where current must be measured.
5. Place the multimeter (ammeter) in series with the circuit at the point where current must be measured.
6. If the multimeter is not autoranging, start on the highest range and switch down to lower ranges until an in-range reading (digital) is obtained (mid- to right-scale reading for greatest accuracy when using an analog multimeter).
7. Stand clear and apply power to measure the current.
8. Remove power to disconnect the multimeter and restore the circuit.

Using The Oscilloscope

APPENDIX F

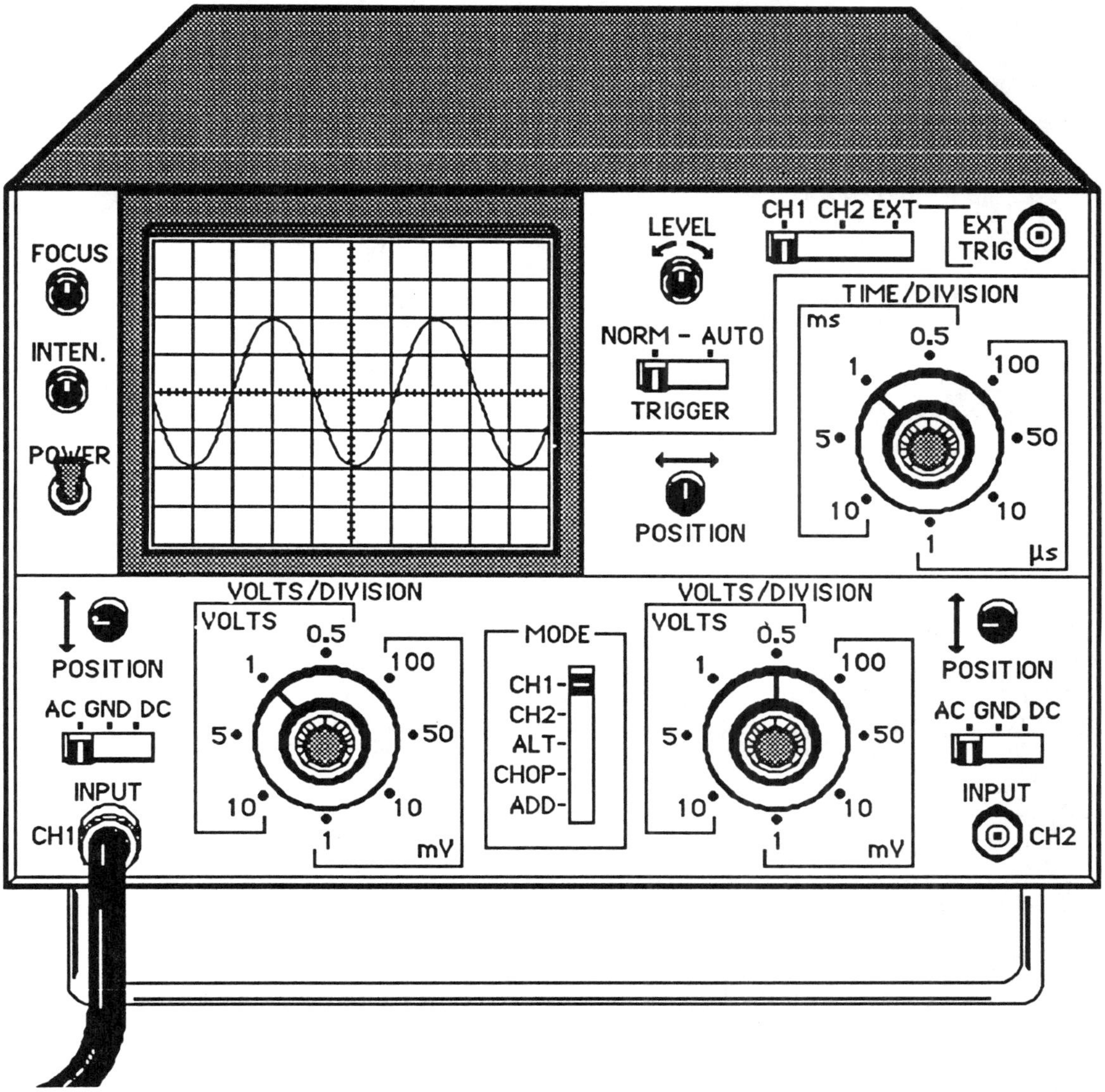

BRIEF DESCRIPTION OF OPERATION

The display screen of an oscilloscope is the face of what is called a cathode ray tube (CRT). All controls on the front of the oscilloscope are used to control the image produced on the screen. The CRT is a vacuum tube that produces a concentrated electron beam from a heated cathode. The electron beam races down the inside center of the CRT at about 1/5 the speed of light. The beam's destination is the back side of the CRT face. The inside face is coated with a phosphor substance that glows (fluoresces) when it is excited by the bombarding electrons and continues to glow (phosphoresce) for a short time after the electron beam has been removed.

As the beam streams down the center of the CRT, it passes through, or near, various electrodes that are used to control the beam's intensity, its focus, and its displacement on the back of the CRT face. The electrodes that are used to control the displacement of the electron beam are called deflection plates. The horizontal deflection plates are used to move the electron beam horizontally from left to right. The vertical deflection plates are used to move the electron beam vertically, up and down.

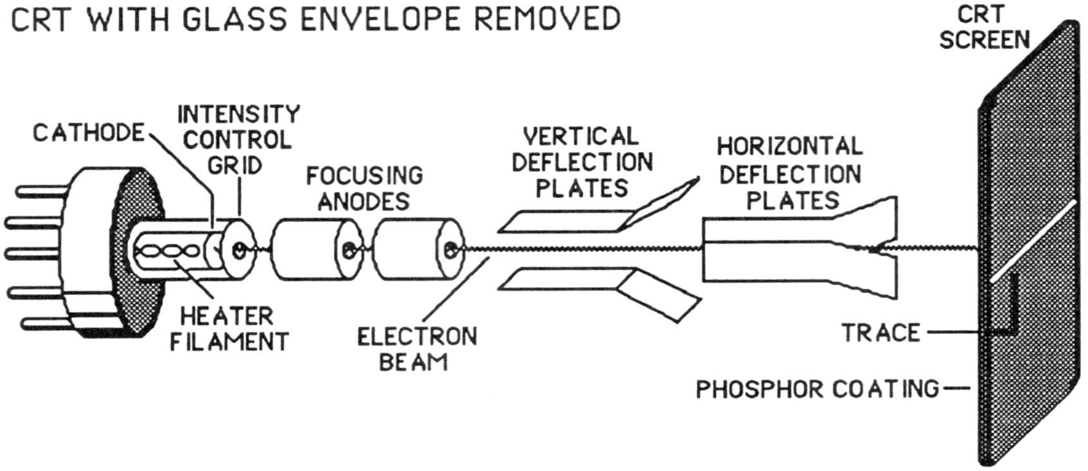

The horizontal deflection plates are energized by the sweep circuitry in the oscilloscope. the sweep circuit consists of a ramp-voltage generator (saw tooth wave) whose frequency can be changed by switching the TIME/DIV control on the front panel. As the ramp voltage increases in voltage, the right deflection plate becomes more positive and the left horizontal deflection plate becomes more negative. This causes the electron beam to sweep from left to right leaving an illuminated trace on the face of the CRT. The length of time it takes for the beam to move from extreme left to extreme right is determined by the cycle time of the ramp voltage, which is set by the TIME/DIV control on the front panel. Thus, the horizontal divisions on the screen represent time.

The vertical deflection plates are energized from a broad-banded amplifier. An external signal is amplified by this amplifier which then drives the vertical deflection plates. This causes the electron beam to deflect vertically. As the electron beam is deflected both horizontally, with sweep time, and vertically, with signal voltage amplitude, a trace is produced that is actually a voltage vs time graph of the applied signal. Therefore, the signal that is being displayed, can be analyzed in terms of peak-to-peak voltage and cycle time (period). A dual-trace oscilloscope can display two waveforms at the same time which makes phase measurements possible. Let's examine the controls of a dual-trace oscilloscope to see how they affect the display.

FRONT PANEL CONTROLS

Power and Display Controls

- Power switch: applies power to the entire oscilloscope.
- Intensity control: used to adjust the brightness of the trace.
- Focus control: used to adjust the sharpness and clarity of the trace.

Vertical Deflection Controls

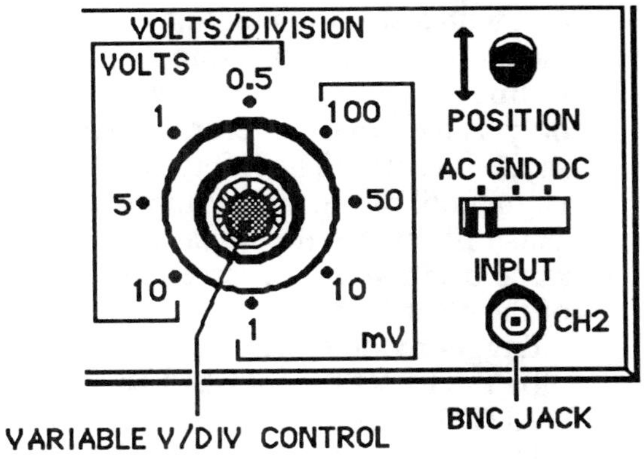

- VOLTS/DIV control: used to assign each major vertical division a voltage. The control shown here is set to scale the screen to 0.5 VOLTS/DIV.
- Variable VOLTS/DIV: used to modify the VOLTS/DIV setting in order to obtain a waveform of a desired size on the screen. This control is normally kept in the fully clockwise calibrate position to insure the accuracy of the main VOLTS/DIV control.
- Position control: used to move the displayed waveform up or down on the screen or to establish a baseline or reference line when measuring DC.
- Input mode selector (AC-GND-DC): In the AC position, only AC is coupled from the input jack to the vertical amplifier (capacitive coupling). In the GND position, the input signal is disconnected and a straight-line trace is displayed indicating the vertical screen position for 0 V or ground. In the DC position, the applied signal is direct coupled to the vertical amplifier. DC and AC are displayed at the same time (AC riding on a measurable DC reference).

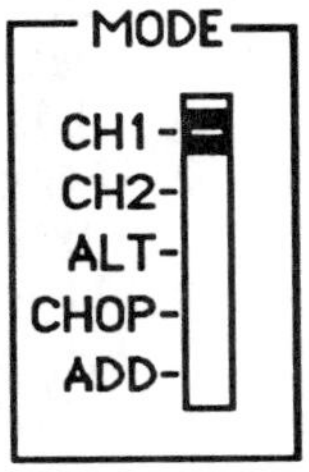

- Display mode selector (CH1-CH2-ALT-CHOP-ADD): CH1 - only channel 1 is displayed, CH2 - only channel 2 is displayed, ALT - both channels are displayed one at a time in very rapid succession so they appear to be displayed at the same time, CHOP - both channels are displayed one small section at a time so the trace jumps back and forth between waveforms as it is moved from left to right across the screen, ADD - the two input signals are added together and the resultant waveform is displayed.

Horizontal Deflection Controls

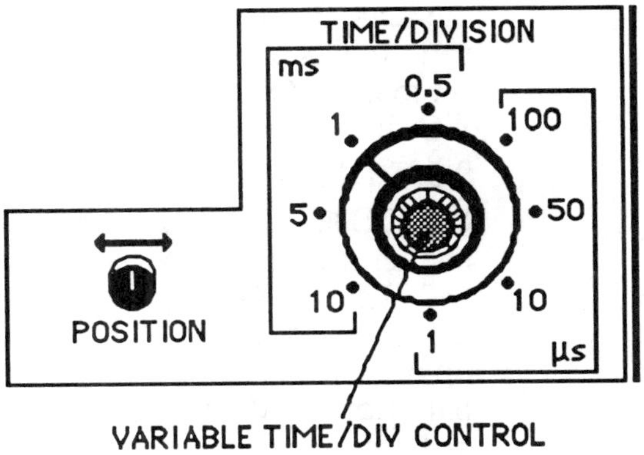

- TIME/DIV control: used to scale the major horizontal divisions to a specific amount of time for each division. The control shown here scales the screen to 1 ms per major horizontal division.
- Variable TIME/DIV control: used to resize the displayed signal horizontally. This control is normally kept in the fully clockwise calibrate position to insure the accuracy of the main TIME/DIV control.
- Horizontal position control: used to move the displayed signal(s) left and right on the screen.

Trigger Controls

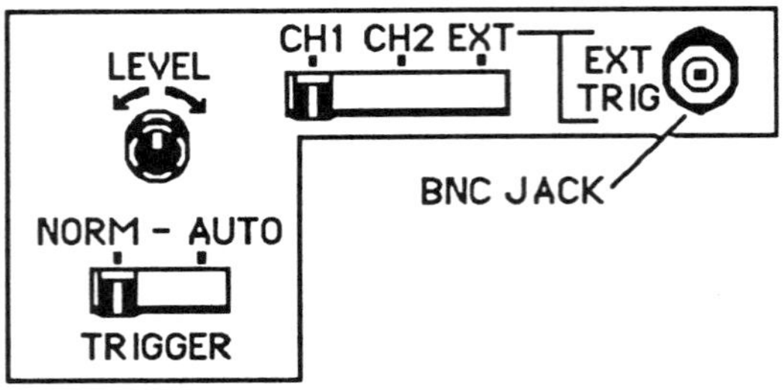

The trigger circuit synchronizes the sweep circuit with a specific point on the waveform of the input signal. For example, the electron beam and sweep circuit may be triggered on only as the input waveform reaches a level of, say, +100 mV. The left to right sweep cannot begin until the +100 mV threshold is reached. This insures that the signal is traced in the same position on the screen with each sweep of the electron beam.

- Trigger mode selector (NORM-AUTO): In the NORM position, the trigger threshold is set by the LEVEL control. Thus, the trigger point can be set by you, the operator. In the AUTO position, the sweep and beam trigger from an internal preset threshold.
- Trigger source selector (CH1-CH2-EXT): The sweep can be triggered from the signal applied to channel 1, channel 2, or from an external signal applied to the EXT TRIG jack.

USING THE OSCILLOSCOPE TO MEASURE VOLTAGE

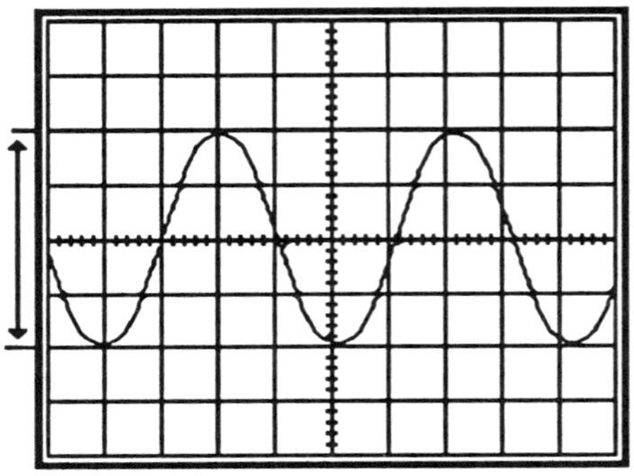

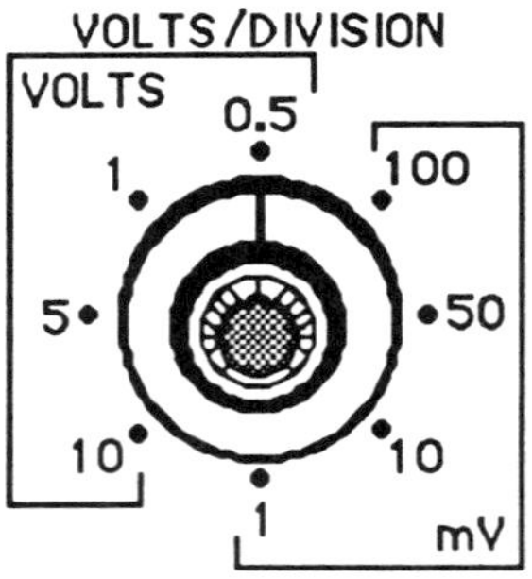

1. Set the TIME/DIV control to display two or three cycles on the screen. (Input mode select set to AC, trigger AUTO from CH1 if channel 1 is being used.)
2. Make sure the variable VOLTS/DIV control is fully clockwise in the calibrate position.
3. Use the vertical position control to place the displayed signal on a reference line so the vertical peak-to-peak distance can be measured. The signal shown here covers 4 major vertical divisions.
4. Multiply the number of vertical divisions by the scale factor set by the VOLTS/DIV control. In this case, the VOLTS/DIV control is set to 0.5 VOLTS/DIV. Thus, the displayed AC signal has an amplitude of 2 $V_{P\text{-}P}$ (4 DIV • 0.5 V/DIV = 2 $V_{P\text{-}P}$).
5. Any DC level can be measured by switching the vertical input mode selector from AC to DC. Use the GND position to establish a reference line then switch to DC to measure the up or down (+ or -) displacement of the signal. Convert the vertical displacement to volts as before.

USING THE OSCILLOSCOPE TO MEASURE CYCLE TIME (PERIOD)

1. Set the TIME/DIV control to display two or three cycles on the screen. (Input mode select set to AC, trigger AUTO from CH1 if channel 1 is being used.)
2. Make sure the variable TIME/DIV control is fully clockwise in the calibrate position.
3. Use the horizontal position control to position the signal so the number of divisions for one cycle can be measured. This could be between postive peaks, negative peaks, or alternate zero-crossing points. In the case shown here, one cycle covers 4.2 horizontal divisions.
4. Multiply the number of horizontal divisions by the scale factor set by the TIME/DIV control. In this case, the TIME/DIV control is set to 1 ms/DIV. Thus, the displayed AC signal has a cycle time (period) of 4.2 ms (4.2 DIV • 1 ms/DIV = 4.2 ms).
5. Calculate the frequency of the signal by taking the reciprocal of its period. In this case, the frequency is 238 Hz (f = 1/T = 1/4.2 ms = 238 Hz).

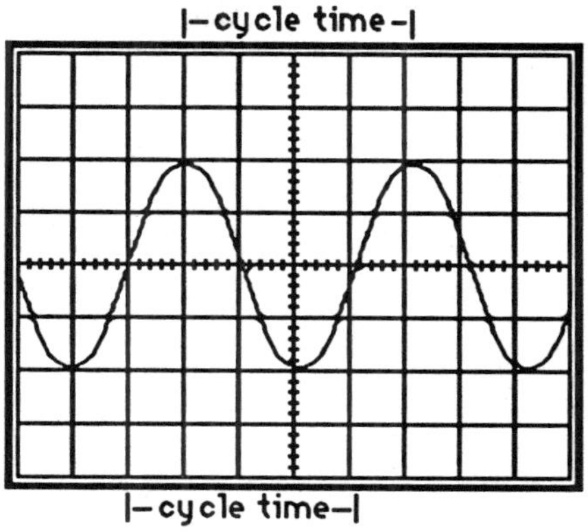

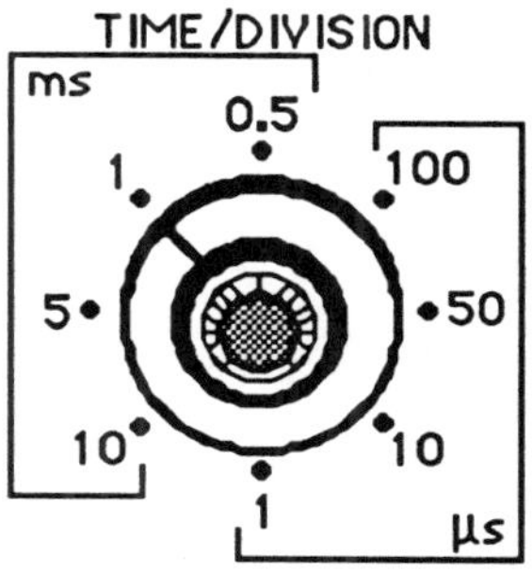

USING THE OSCILLOSCOPE TO MEASURE THE TIME DIFFERENCE AND PHASE SHIFT BETWEEN TWO SIGNALS

1. Measure the time for one cycle, in this case, 4.2 ms per cycle.
2. Measure the time difference, or delay, between like zero-crossing points or between positive or negative peaks. In the case shown, the delay is 1 ms.
3. Calculate the phase shift (time expressed in degrees) using this formula: phase angle = 360° • time delay/cycle time. In this case, the phase shift is 85.7° (360° • 1 ms/4.2 ms = 85.7°).

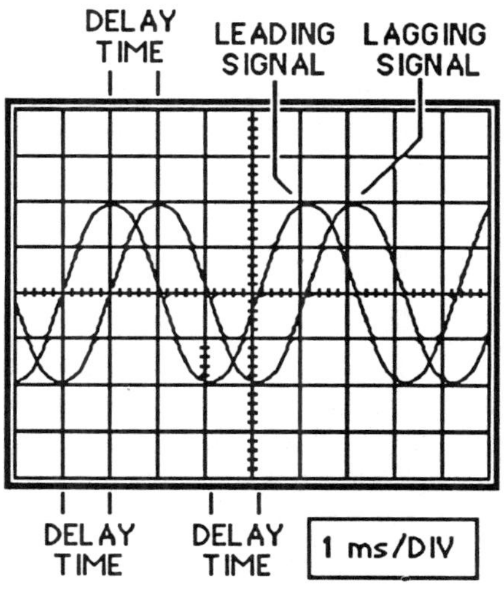

Soldering Techniques

APPENDIX G

SOLDER = LEAD + TIN

Solder is a combination of lead and tin that ranges from 100% lead to 100% tin. The mix of lead and tin determines important solder characteristics such as: melting point, plastic state, and wetting quality.

The **melting point** is the temperature at which the solder leaves the solid state. Pure tin melts at 450 ° F and pure lead melts at 621° F. Combining lead and tin actually lowers the melting point to as low as 361° F for a 63% tin, 37% lead, solder mixture.

The **plastic state** is the soft melted state that exists between the solid and liquid states as the solder is heated or cooled. Pure lead and pure tin have no plastic state since the transition from solid to liquid is immediate at the melting point. However, all mixtures of lead and tin have a plastic state except a ratio of 63% tin to 37% lead. Thus, 63/37 solder changes quickly from solid to liquid, and vice versa, at about 361° F. The plastic state of solders is undesireable for applications where mechanical movement or vibration may occur as the solder cools. Any disturbance of the solder during the plastic state causes imperfections and deformaties to remain in the solder. The underminds the mechanical and electrical quality of the solder creating what is called a **cold solder joint** (a poor connection with high resistance).

The **wetting quality** of solder is the ability of the solder to spread and bond quickly when it reaches its liquid state. This bonding is usually the creation of an alloy of solder and metal (usually copper). Thus, the alloy forms a very strong and low-resistance bridge between the solder and the metal surface.

Because of its low melting point, very short plastic state, and very good wetting quality, **60/40 solder** is the most common solder used in electronics (60% tin/40% lead).

SOLDER FLUX

Flux is a deoxidizing agent used in the soldering process. Metal surfaces oxidize quickly in the air. This means the metal surface forms molecular bonds with oxygen. The oxidation inhibits the bonding of the solder to the metal surface. When flux is used during the soldering process, the oxidation is removed and floated away. The presence of the flux also prevents further oxidation but does not interfere with the bonding of the solder to the metal surface.

Rosin core solder is the most common solder used in electronics. Rosin is a noncorrosive flux that is manufactured into the core of the solder. The rosin is actually placed in a small tunnel down the center of the solder. During the soldering process, the right amount of flux is automatically applied to the metal surfaces before the solder reaches the liquid state. 60/40 rosin core solder is the most common and is available in many diameters. Be careful not to purchase acid-core solder for electrical work. The acid flux will destroy wiring and copper foils on PC boards.

Flux is not a cleaning agent. It cannot be expected to clear away large amounts of oxidation and corrosion. Surfaces that are not shiny and visibly clean must be cleaned with a fine steelwool or emery cloth. The oxidation that instantly returns, after manual cleaning, is easily taken care of by the flux during the soldering process.

SOLDERING EQUIPMENT

There is a vast selection of soldering equipment. Here is a brief list of common equipment and applications.

- Heavy duty 500 to 2000 W soldering irons used to heat large metal surfaces such as chassis pieces and large cable end terminals.
- 100 to 200 W soldering guns used for chassis point-to-point wiring involving terminal lugs and relatively large-gauge wire (18 to 12 gauge wire).
- Low wattage soldering irons (40 to 60 W) used for light duty point-to-point wiring and printed circuit board soldering.
- Very low-wattage pencil soldering irons (20 to 40 W) for very light duty printed circuit board work (for mounting transistors and integrated circuits, etc.).
- Rechargeable cordless soldering irons for light duty point-to-point wiring and printed circuit board work where a small number of solder joints must be made (in most cases, less than 100 joints per charge).
- Temperature controlled soldering stations used for delicate components such as integrated circuits, small diodes, and transistors.

POINT-TO-POINT TERMINAL SOLDERING

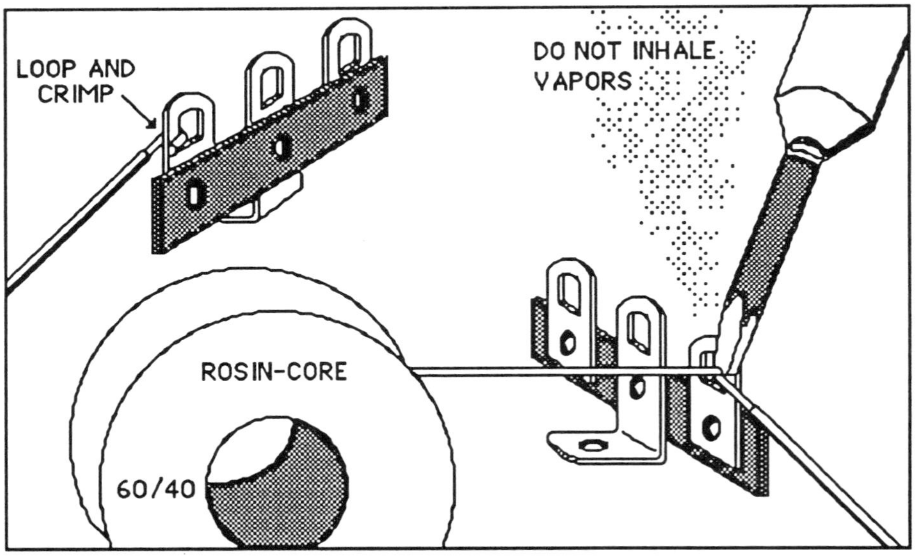

- Use a 100 to 200 W soldering gun or a 40 to 50 W soldering iron.
- Use 60/40 rosin core solder.
- Clean the terminal and wire(s) before soldering if necessary.
- The preheated soldering tip must be placed in contact with the terminal surface and the wire at the same time.
- Apply a little solder at the point of contact between the soldering iron tip and the wire. Then, apply solder to the opposite side of the wire and terminal. The solder should flow quickly and smoothly all around the wire and terminal contact area. If it does not, the iron is not hot enough and/or the surfaces are not clean.
- Do not disturb the solder joint as it cools to avoid a cold solder joint.

PRINTED CIRCUIT BOARD SOLDERING

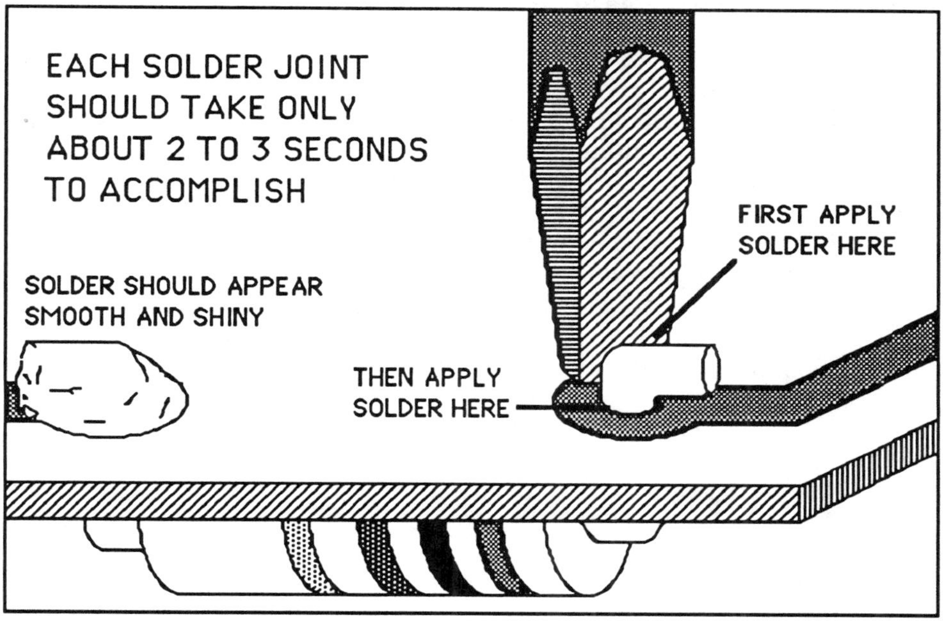

- Make sure the copper foil and component leads are clean.
- Mount the component flush against the component side of the PC board.
- Bend the component leads flat against the foil side of the PC board to hold the component in place.
- Clip off any length of component lead that exceeds the perimeter of the foil pad.
- Use a low-wattage soldering iron, or pencil iron (20 to 60 W).
- Make sure the iron is fully heated. Never allow the iron to heat up while holding the tip against the foil or component lead.
- Place the iron tip against the foil and component lead and immediately apply a small amount of solder between the tip and wire lead, then wire and foil. Work quickly so the foil and component are not damaged by overheating. The solder joint should be smooth, shiny, and cover only the wire and foil pad.
- When soldering small delicate components such as small diodes, transistors, and integrated circuits, you must work quickly. Applying heat too long will damage these devices. It should not take any longer than 2 or 3 seconds to make a good solder joint.
- When soldering integrated circuit pins to the foil, butt the iron tip against the foil and pin at the same time. Apply solder to the mating point of the tip, pin, and foil.

GENERAL SOLDERING HINTS

- Do not breath the vapors that are released into the air during soldering. Solder in a well ventilated area.
- Keep the soldering iron tip clean and shiny by frequently wiping it across a moist cloth or sponge (not a synthetic sponge).
- Do not allow excess solder to buildup on the tip. Wipe it off.
- Do not balance a hot iron on the edge of your work table. Use a soldering iron holder to secure the iron when not in use.
- Use rubbing alcohol to clean flux residue away from solder joints when finished. Use cotton swabs to apply or lint-free paper towels.
- Keep your work area clean and organized to permit quality work.
- Do not leave the iron on any longer than you need it. This will increase the life of the soldering tip.

Wire-Wrapping

APPENDIX **H**

Wire-wrapping is a practical alternative to soldering for wiring anything from large systems to small circuits. The telephone industry has used wire-wrap technology for many years and has proven it to be reliable for millions of telephone connections. In the past 15 to 20 years, wire-wrapping has become popular for studio wiring and electronic circuit prototyping. Wire-wrapping has become popular because it is quick, clean, can easily be undone and redone, and there is no need for a hot iron.

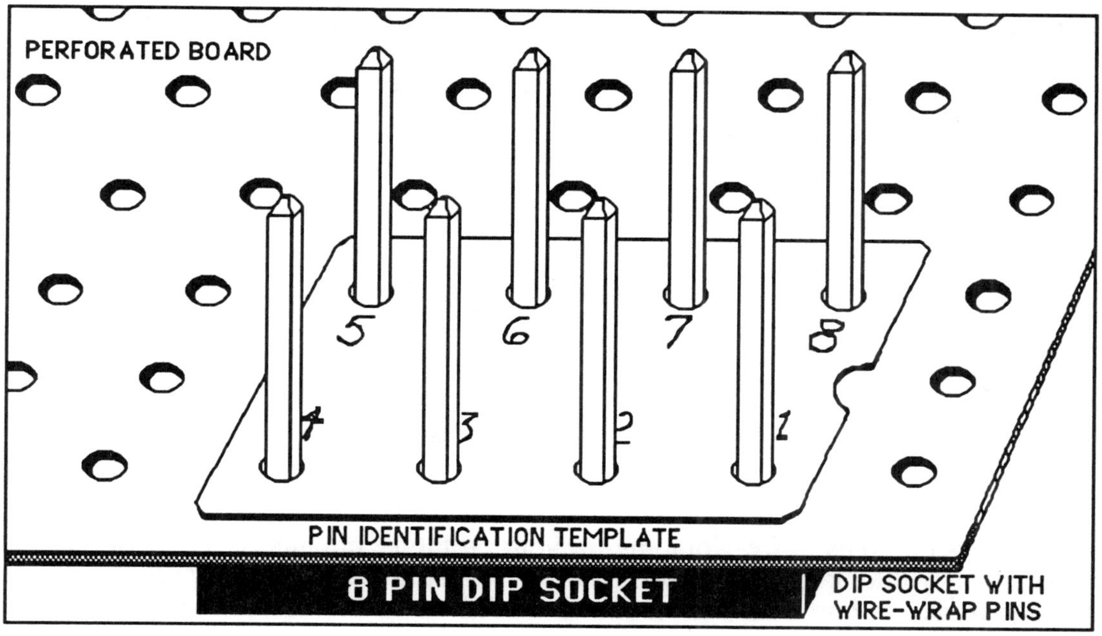

As shown in the diagram above, integrated circuit sockets are available that have long square wire-wrap pins. Individual terminals and header plugs are also available for mounting individual components such as transistors, resistors, capacitors, diodes, etc.. As you can see, a perforated board is used to mount the wire-wrap sockets and terminals. The hole spacing for such a board is a standard 0.1 inch, matching the IC pin spacing. Pin identification templates can be used to simplify the wiring process and reduce errors.

Why are wire-wrap pins square instead of round? The square corners of the pins become pressure points that secure a very tight wrap. The wire is forced to make an airtight contact against the corner edges. This insures a good corrosion free contact for many years. If the terminal was round, the wire tension would be evenly distributed forming a lose wrap.

Standard AWG wire gauges used for wire-wrapping range from #20 down to #32. The wire-wrap tool must be selected to match the gauge of wire that must be used. Some wire-wrap tools have interchangeable bits that can be changed to accommodate the desired gauge wire. The tool is designed to wrap the wire very securely around the post as the wire feed bit is turned by hand or motor. The following diagram illustrates the kind of results you want for each connection. There

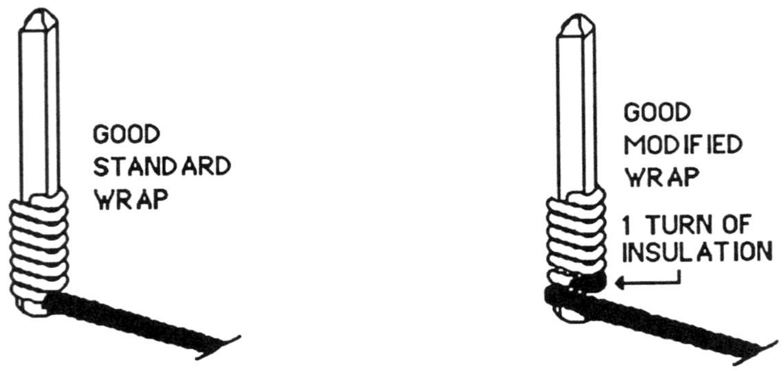

should be 6 to 10 tight closely wound turns on the post for each connection. A little practice is all it takes to make perfect connections. The modified wrap illustrated above is used for very small gauge wire (#30 & #32) and permits the wire's insulation to contribute to the mechanical strength of the connection.

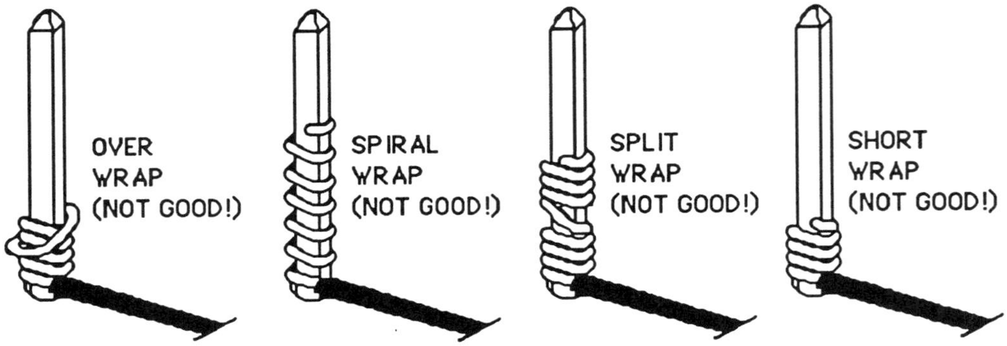

When you first begin to use wire-wrapping for your prototype circuits, you will find that some of your connections will look like those shown above. These are unsatisfactory connections that can become undermined by corrosion and mechanical vibration. Do not allow yourself to accept this kind of work. Troubleshooting a wire-wrapped circuit for a poor wire-wrap connection can be a real challenge. Again, make sure each connection is 6 to 10 turns of tight closely wrapped wire. Each post, or terminal, should accommodate 2 or 3 connections.

There is a wide range of wire-wrap tools from which you can choose. They range from electric guns with automatic stripping and wire feed to a simple combination hand tool as shown below. For small projects, the combination hand tool is very adequate. As stated earlier, the tool must be selected to handle the gauge wire that is needed for your project. Hand wrap tools are available for

Wire-Wrap Combination Hand Tool

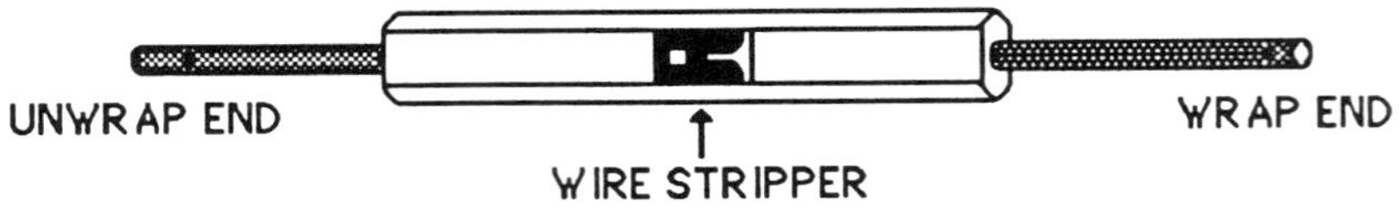

all wire sizes from #20 to #32. Usually, #30 or #32 wire is used for small project breadboarding (prototyping) and #22 to #26 gauge wire is used for large system wiring (studio wiring, etc.).

Equipment and Parts

APPENDIX I

Exploring Electronic Devices Laboratory Manual

EQUIPMENT

- ❑ 1 - Dual-trace oscilloscope
- ❑ 1 - Bench function generator (to 2 MHz)
- ❑ 2 - Clip-lead oscilloscope cables
- ❑ 1 - Analog FET VOM or DMM
- ❑ 1 - DC supply (dual 0 to 25 V)
- ❑ 1 - Set of generator clip-lead cables

PARTS NEEDED FOR EXPERIMENTS 1 THROUGH 22

RESISTORS

Trimmer Potentiometers, Wire Leads, Protoboard Mountable, Thumbwheel

❑ 1 - 1 kΩ ❑ 1 - 5 kΩ ❑ 3 - 10 kΩ ❑ 1 - 50 kΩ

1/4 W, 5% Film Resistors

- ❑ 1 - 100 kΩ
- ❑ 2 - 47 kΩ
- ❑ 1 - 30 kΩ
- ❑ 1 - 22 kΩ
- ❑ 1 - 18 kΩ
- ❑ 2 - 4.7 kΩ
- ❑ 1 - 2.2 kΩ
- ❑ 2 - 1 kΩ
- ❑ 1 - 470 Ω
- ❑ 1 - 270 Ω
- ❑ 1 - 150 Ω
- ❑ 2 - 100 Ω
- ❑ 1 - 22 Ω
- ❑ 2 - 68 kΩ
- ❑ 1 - 43 kΩ
- ❑ 1 - 27 kΩ
- ❑ 1 - 20 kΩ
- ❑ 2 - 10 kΩ
- ❑ 1 - 3.3 kΩ
- ❑ 1 - 1.5 kΩ
- ❑ 1 - 680 Ω
- ❑ 1 - 330 Ω
- ❑ 1 - 220 Ω
- ❑ 1 - 120 Ω
- ❑ 1 - 33 Ω
- ❑ 1 - 10 Ω

Power Resistors

- ❑ 1 - 1 kΩ, 1 W, 5%
- ❑ 1 - 220 Ω, 2 W, 5%
- ❑ 1 - 100 Ω, 2 W, 5%
- ❑ 1 - 330 Ω, 1 W, 5%
- ❑ 1 - 120 Ω, 1 W, 5%
- ❑ 1 - 51 Ω, 1/2 W, 5%

CAPACITORS

- ❑ 1 - 1,000 μF, 35 V Electrolytic
- ❑ 2 - 10 μF, 35 V Electrolytic
- ❑ 1 - 0.1 μF, 100 V Film
- ❑ 1 - 0.001 μF, 100 V Film
- ❑ 3 - 100 μF, 50 V Electrolytic
- ❑ 2 - 1 μF, 35 V Electrolytic
- ❑ 1 - 0.01 μF, 100 V Film
- ❑ 1 - 100 pF, Dipped Mica

INDUCTORS

- ❑ 1 - 1 mH High-Q
- ❑ 1 - 115 V/25.2 V CT, 450 to 500 mA Transformer with safely attached power cord, preferably in an insulated case with banana-jack terminals for the secondary winding.

DIODES

- ❑ 3 - 1N914 or 1N4148
- ❑ 1 - 1N4742 Zener, 12 V, 1 W
- ❑ 1 - Red or green LED
- ❑ 4 - 1N4001 Diodes, 50 V, 1 A
- ❑ 1 - 1N4733 Zener, 5.1 V, 1 W

TRANSISTORS

- ❑ 3 - 2N3904 NPN, General purpose
- ❑ 2 - 2N2219 NPN Medium-power
- ❑ 1 - IRF511 Power MOSFET with heat sink
- ❑ 2 - 2N3906 PNP, General purpose
- ❑ 2 - MPF102 N-Channel JFET

MISCELLANEOUS MATERIALS

- ❑ Misc. lengths of solid #22 insulated hookup wire (red, black, white)
- ❑ 1 - 6" Plugin protoboard (breadboard)

ADDITIONAL PARTS NEEDED FOR EXPERIMENTS 23 THROUGH 44

RESISTORS

- ❑ 1 - 100 kΩ, Trimmer, Protoboard Mount
- ❑ 1 - 330 kΩ, 1/4 W Film, 5%
- ❑ 1 - 56 kΩ, 1/4 W Film, 5%
- ❑ 3 - 20 kΩ, 1/4 W Film, 5%
- ❑ 1 - 6.8 kΩ, 1/4 W Film, 5%
- ❑ 1 - 2 kΩ, 1/4 W Film, 5%
- ❑ 1 - 51 Ω, 1 W Film, 5%
- ❑ 1 - 5.9 kΩ, 1/4 W Film, 2%
- ❑ 1 - 1 MΩ, 1/4 W Film, 5%
- ❑ 2 - 100 kΩ, 1/4 W Film, 5%
- ❑ 1 - 33 kΩ, 1/4 W film resistor, 5%
- ❑ 2 - 10 kΩ, 1/4 W Film, 5%
- ❑ 1 - 2.2 kΩ, 1/4 W Film, 5%
- ❑ 1 - 2 kΩ, 1 W Film, 5%
- ❑ 1 - 20 Ω, 1/4 W Film, 5%
- ❑ 1 - 10 Ω, 1/4 W Film, 2%

CAPACITORS

- ❑ 1 - 0.1 μF, 100 V Film
- ❑ 1 - 0.01 μF, 100 V Film

INTEGRATED CIRCUITS, THYRISTORS and MISC.

- ❑ 1 - LM741 Op-amp
- ❑ 1 - LM555 Timer
- ❑ 1 - LM723 or μA723 Regulator
- ❑ 1 - 2N6342 or 2N6343 TRIAC
- ❑ 1 - MOC3010 TRIAC Optoisolator
- ❑ 3 - LF356 Op-amps
- ❑ 1 - XR2206 Function generator
- ❑ 1 - 2N1595 or 2N1596 SCR
- ❑ 1 - 2N4851 UJT
- ❑ 1 - #1815 Lamp, 14 V, 0.2 A